Excel for Statistics

Excel for Statistics is a series of textbooks that explain how to use Excel to solve statistics problems in various fields of study. Professors, researchers, students, and practitioners will find these books teach how to make Excel work best in their respective field. Applications include any discipline that uses data and can benefit from the power and simplicity of Excel. Books cover all the steps for running statistical analyses in Excel 2019, Excel 2016, and Excel 2013. The approach also teaches critical statistics skills, making the books particularly applicable for statistics courses taught outside of mathematics or statistics departments.

Series editor: Thomas J. Quirk

The following books are in this series:

T.J. Quirk, *Excel 2016 for Engineering Statistics: A Guide to Solving Practical Problems*, Excel for Statistics. Springer International Publishing Switzerland 2016.

T.J. Quirk, *Excel 2016 for Business Statistics: A Guide to Solving Practical Problems*, Excel for Statistics. Springer International Publishing Switzerland 2016.

T.J. Quirk, M. Quirk, H.F. Horton, *Excel 2016 for Biological and Life Sciences Statistics: A Guide to Solving Practical Problems*, Excel for Statistics. Springer International Publishing 2016.

T.J. Quirk, M. Quirk, H.F. Horton, *Excel 2016 for Physical Sciences Statistics: A Guide to Solving Practical Problems*, Excel for Statistics. Springer International Publishing 2016.

T.J. Quirk, E. Rhiney, *Excel 2016 for Marketing Statistics: A Guide to Solving Practical Problems*, Excel for Statistics. Springer International Publishing Switzerland 2016.

T.J. Quirk, *Excel 2016 for Educational and Psychological Statistics: A Guide to Solving Practical Problems*, Excel for Statistics. Springer International Publishing Switzerland 2016.

T.J. Quirk, *Excel 2016 for Social Science Statistics: A Guide to Solving Practical Problems*, Excel for Statistics. Springer International Publishing Switzerland 2016.

T.J. Quirk, S. Cummings, *Excel 2016 for Health Services Management Statistics: A Guide to Solving Practical Problems*, Excel for Statistics. Springer International Publishing Switzerland 2016.

T.J. Quirk, J. Palmer-Schuyler, *Excel 2016 for Human Resource Management Statistics: A Guide to Solving Practical Problems*, Excel for Statistics. Springer International Publishing Switzerland 2016.

T.J. Quirk, M. Quirk, H.F. Horton. *Excel 2016 for Environmental Sciences Statistics: A Guide to Solving Practical Problems*, Excel for Statistics. Springer International Publishing Switzerland 2016.

T.J. Quirk, J. Palmer-Schuyler, *Excel 2013 for Human Resource Management Statistics: A Guide to Solving Practical Problems*, Excel for Statistics. Springer International Publishing Switzerland 2016.

T.J. Quirk, S. Cummings, *Excel 2013 for Health Services Management Statistics: A Guide to Solving Practical Problems*. Excel for Statistics. Springer International Publishing Switzerland 2016.

T.J. Quirk, M. Quirk, H.F. Horton, *Excel 2013 for Physical Sciences Statistics: A Guide to Solving Practical Problems*. Excel for Statistics. Springer International Publishing Switzerland 2016.

T.J. Quirk. *Excel 2013 for Educational and Psychological Statistics: A Guide to Solving Practical Problems*, Excel for Statistics. Springer International Publishing Switzerland 2015.

T.J. Quirk, M. Quirk, H.F. Horton, *Excel 2013 for Biological and Life Sciences Statistics: A Guide to Solving Practical Problems*, Excel for Statistics. Springer International Publishing Switzerland 2015.

T.J. Quirk, *Excel 2013 for Social Science Statistics: A Guide to Solving Practical Problems*, Excel for Statistics. Springer International Publishing Switzerland 2015.

T.J. Quirk, *Excel 2013 for Business Statistics: A Guide to Solving Practical Problems*, Excel for Statistics. Springer International Publishing Switzerland 2015.

T.J. Quirk. *Excel 2013 for Engineering Statistics: A Guide to Solving Practical Problems*, Excel for Statistics. Springer International Publishing Switzerland 2015.

T.J. Quirk, M. Quirk, H.F. Horton, *Excel 2013 for Environmental Sciences Statistics: A Guide to Solving Practical Problems*, Excel for Statistics. Springer International Publishing Switzerland 2015.

Additional Statistics books by Dr. Tom Quirk that have been published by Springer

T.J. Quirk, *Excel 2010 for Engineering Statistics: A Guide to Solving Practical Problems*. Springer International Publishing Switzerland 2014.

T.J. Quirk, S. Cummings, *Excel 2010 for Health Services Management Statistics: A Guide to Solving Practical Problems*. Springer International Publishing Switzerland 2014.

T.J. Quirk, M. Quirk, H. Horton, *Excel 2010 for Physical Sciences Statistics: A Guide to Solving Practical Problems*. Springer International Publishing Switzerland 2013.

T.J. Quirk, M. Quirk, H.F. Horton, *Excel 2010 for Biological and Life Sciences Statistics: A Guide to Solving Practical Problems*. Springer Science+Business Media New York 2013.

T.J. Quirk, M. Quirk, H.F. Horton, *Excel 2007 for Biological and Life Sciences Statistics: A Guide to Solving Practical Problems*. Springer Science+Business Media New York 2013.

T.J. Quirk, *Excel 2010 for Social Science Statistics: A Guide to Solving Practical Problems*. Springer Science+Business Media New York 2012.

T.J. Quirk, *Excel 2010 for Educational and Psychological Statistics: A Guide to Solving Practical Problems*. Springer Science+Business Media New York 2012.

T.J. Quirk, *Excel 2007 for Business Statistics: A Guide to Solving Practical Problems*. Springer Science+Business Media New York 2012.

T.J. Quirk, *Excel 2007 for Social Science Statistics: A Guide to Solving Practical Problems*. Springer Science+Business Media New York 2012.

T.J. Quirk, *Excel 2007 for Educational and Psychological Statistics: A Guide to Solving Practical Problems*. Springer Science+Business Media New York 2012.

More information about this series at http://www.springer.com/series/13491

Thomas J. Quirk • Meghan H. Quirk
Howard F. Horton

Excel 2019 for Environmental Sciences Statistics

A Guide to Solving Practical Problems

Second Edition

 Springer

Thomas J. Quirk
Professor Emeritus of Marketing,
Webster University
St. Louis, MO, USA

Meghan H. Quirk
Thornton, CO, USA

Howard F. Horton
Thornton, CO, USA

ISSN 2570-4605   ISSN 2570-4613   (electronic)
Excel for Statistics
ISBN 978-3-030-66276-9   ISBN 978-3-030-66277-6   (eBook)
https://doi.org/10.1007/978-3-030-66277-6

This Springer imprint is published by the registered company Springer Nature Switzerland AG
The registered company address is: Gewerbestrasse 11, 6330 Cham, Switzerland

This book is dedicated to the more than 3,000 students I have taught at Webster University's campuses in St. Louis, London, and Vienna; the students at Principia College in Elsah, Illinois; and the students at the Cooperative State University of Baden-Wuerttemberg in Heidenheim, Germany. These students taught me a great deal about the art of teaching. I salute them all, and I thank them for helping me to become a better teacher.

Thomas J. Quirk

We dedicate this book to all the newly inspired students emerging into the ranks of the various fields of science.

Meghan H. Quirk and Howard F. Horton

Preface

Excel 2019 for Environmental Sciences Statistics: A Guide to Solving Practical Problems updates the Excel steps and screenshots from the previously published *Excel 2016 for Environmental Sciences Statistics: A Guide to Solving Practical Problems*, and it contains a number of important changes. The explanations of statistics and statistical formulas have been made clearer. The Excel steps now match perfectly the Excel 2019 version. Thirty percent of the end-of-chapter problems, and their answers in an Appendix, are new to this book. Thirty percent of the 160+ screenshots are new so that they match the new Excel commands to ensure that you are using Excel correctly each step of the way.

The eight chapters in the book (Mean, Standard Deviation, and Standard Error of the Mean; Random Sampling; Confidence Interval about the Mean; One-Group t-test; Two Group t-test; Correlation and Linear Regression; Multiple Correlation; and One-Way Analysis of Variance) have been rewritten to improve their explanation of statistics. The answers to all of the problems in the book are provided, and there is a Practice Test so that you can test your ability to solve statistics problems using Excel. This book is an introduction to statistics, not a full-blown explanation of statistics.

A word of caution: This book does not attempt to teach you all of the "bells and whistles" of Excel 2019. We have left that objective to other books. Instead, this book will teach you the Excel steps you need to solve the interesting problems in the book. You should think of Excel as merely the "computer language" needed to solve statistics problems. In a sense, this approach is similar to the one you would need if you planned to spend a year living in Europe in Vienna, Austria, where you needed to learn some basic German (e.g., "How much does this cost?" "Where is the train station?" "Please give me the bill for my dinner." "How can I get to the airport?"), but you do not need to become fluent in that language to survive. This book focuses on the Excel steps needed to solve the problems in the book. The task of showing you how to use the many powers of Excel are beyond the scope of this book.

This book was written by a Professor who wanted to respond to the complaints of many students about their inability to understand their statistics textbook and about their inability to understand their professor's explanation of theoretical statistics. This book is self-instructional and does not depend on a professor's explanation of statistics. This book will teach you the general concepts of statistics without burying you in dull statistical theory. You will learn *why* you are performing the Excel steps through the objectives included in the chapters. The statistical concepts and practice problems get progressively more sophisticated as they build on what you have already learned from studying this book. This book is understandable by both undergraduate and graduate students who are taking their first course in statistics, by researchers, and by working professionals who want to solve interesting problems in their chosen field of study.

This book was written by a Professor who is, first and foremost, committed to helping you to understand how to use statistics to solve interesting problems in your chosen field of study. The ideas in this book have been classroom tested over the past 11 years in both undergraduate and graduate courses at Webster University, a liberal arts college located in St. Louis, Missouri, in the middle of the USA. This book is part of a series of more than 30 introductory statistics textbooks, in 12 fields of study, that have been published by Springer by Prof. Quirk, which have helped thousands of students, researchers, and working professionals learn how to use Excel to solve interesting statistics problems. These fields of study include: (1) Business, (2) Education/Psychology, (3) Social Science, (4) Biological and Life Sciences, (5) Physical Sciences, (6) Engineering, (7) Health Services Management, (8) Human Resource Management, (9) Environmental Sciences, (10) Marketing, (11) Social Work, and (12) Advertising.

Thomas J. Quirk, a current Professor Emeritus of Marketing at The Walker School of Business & Technology at Webster University in St. Louis, Missouri (USA), taught Marketing Statistics, Marketing Research, and Pricing Strategies. He has published articles in *The Journal of Educational Psychology, Journal of Educational Research, Review of Educational Research, Journal of Educational Measurement,* and *Educational Technology.* Professor Quirk has published more than 20 articles in professional journals, and presented more than 20 papers at professional meetings. He holds a BS in Mathematics from John Carroll University, both an MA in Education and a PhD in Educational Psychology from Stanford University, and an MBA from the University of Missouri-St. Louis.

Meghan H. Quirk holds both a PhD in Biological Education and an MA in Biological Sciences from the University of Northern Colorado (UNC), and a BA in Biology and Religion at Principia College in Elsah, Illinois. She has done research on food web dynamics at Wind Cave National Park in South Dakota and research in agro-ecology in Southern Belize. She has coauthored an article on shortgrass steppe ecosystems in *Photochemistry & Photobiology* and has presented papers at the Shortgrass Steppe Symposium in Fort Collins, Colorado, the Long-term Ecological Research All Scientists Meeting in Estes Park, Colorado, and participated in the NSF Site Review of the Shortgrass Steppe Long Term Ecological Research in Nunn, Colorado. She was a National Science Foundation Fellow GK-12, and currently is a public school administrator in Erie, Colorado.

Howard F. Horton holds an MS in Biological Sciences from the University of Northern Colorado (UNC) and a BS in Biological Sciences from Mesa State College. He has worked on research projects in Pawnee National Grasslands, Rocky Mountain National Park, Long Term Ecological Research at Toolik Lake, Alaska, and Wind Cave, South Dakota. He has co-authored articles in *The International Journal of Speleology* and *The Journal of Cave and Karst Studies.* He was a National Science Foundation Fellow GK-12, and a District Wildlife Manager with the Colorado Division of Parks and Wildlife. He is currently Organizational Development Coordinator with the Colorado Parks and Wildlife (USA).

St. Louis, MO, USA   Thomas J. Quirk
Thornton, CO, USA   Meghan H. Quirk
                                                                                    Howard F. Horton

Acknowledgments

Excel 2019 for Environmental Sciences Statistics: A Guide to Solving Practical Problems is the result of inspiration from three important people: my two daughters and my wife. Jennifer Quirk McLaughlin invited me to visit her MBA classes several times at the University of Witwatersrand in Johannesburg, South Africa. These visits to a first-rate MBA program convinced me there was a need for a book to teach students how to solve practical problems using Excel. Meghan Quirk-Horton's dogged dedication to learning the many statistical techniques needed to complete her PhD dissertation illustrated the need for a statistics book that would make this daunting task more user-friendly. And Lynne Buckley-Quirk was the number-one cheerleader for this project from the beginning, always encouraging me and helping me remain dedicated to completing it.

Thomas J. Quirk

We would like to acknowledge the patience of our two little girls, Lila and Elia, as we worked on this book with their TQ. We would also like to thank Professors Sarah Perkins, Doug Warren, John Moore, and Lee Dyer for their guidance and support during our college and graduate school careers.

Meghan H. Quirk and Howard F. Horton

Contents

Chapter 1
Sample Size, Mean, Standard Deviation, and Standard Error of the Mean

There are many forces in our environment on this earth that place our environment and people at risk. Problems such as acid rain, climate change, deforestation, greenhouse gases, pollution of streams and rivers, emissions from diesel fuel, disposal of nuclear waste, carbon emissions, and so forth are just some of the serious challenges faced by people on earth.

Statistics is one way of studying these types of problems in order to understand better how our attempts to reduce the harmful effects of these problems can improve our environment. Environmental sciences is a multidisciplinary field of study that includes biology, ecology, chemistry, physics, mineralogy, soil science, geology, engineering, natural resource management, air and noise pollution control, climate change, and geography. As such, the study of environmental sciences cuts across many disciplines and fields of study.

This chapter deals with how you can use Excel to find the average (i.e., "mean") of a set of scores, the standard deviation of these scores (STDEV), and the standard error of the mean (s.e.) of these scores. All three of these statistics are used frequently and form the basis for additional statistical tests.

1.1 Mean

The *mean* is the "arithmetic average" of a set of scores. When my daughter was in the fifth grade, she came home from school with a sad face and said that she didn't get "averages." The book she was using described how to find the mean of a set of scores, and so I said to her:

"Jennifer, you add up all the scores and divide by the number of numbers that you have."
She gave me "that look," and said: "Dad, this is serious!" She thought I was teasing her.
So I said:
"See these numbers in your book; add them up. What is the answer?" (She did that.)
"Now, how many numbers do you have?" (She answered that question.)

© Springer Nature Switzerland AG 2021
T. J. Quirk et al., *Excel 2019 for Environmental Sciences Statistics*,
Excel for Statistics, https://doi.org/10.1007/978-3-030-66277-6_1

"Then, take the number you got when you added up the numbers, and divide that number by the number of numbers that you have."

She did that and found the correct answer. You will use that same reasoning now, but it will be much easier for you because Excel will do all of the steps for you.

We will call this average of the scores the "mean" which we will symbolize as: $\overline{X}$, and we will pronounce it as: "Xbar."

The formula for finding the mean with your calculator looks like this:

$$\overline{X} = \frac{\Sigma X}{n} \tag{1.1}$$

The symbol Σ is the Greek letter sigma, which stands for "sum." It tells you to add up all the scores that are indicated by the letter X, and then to divide your answer by n (the number of numbers that you have).

Let's give a simple example:

Suppose that you had these six environmental science test scores on a seven-item true-false quiz:

6
4
5
3
2
5

To find the mean of these scores, you add them up, and then divide by the number of scores. So, the mean is: $25/6 = 4.17$.

1.2 Standard Deviation

The *standard deviation* tells you "how close the scores are to the mean." If the standard deviation is a small number, this tells you that the scores are "bunched together" close to the mean. If the standard deviation is a large number, this tells you that the scores are "spread out" a greater distance from the mean. The formula for the standard deviation (which we will call STDEV) and use the letter, S, to symbolize is:

$$\text{STDEV} = S = \sqrt{\frac{\Sigma(X - \overline{X})^2}{n - 1}} \tag{1.2}$$

The formula looks complicated, but what it asks you to do is this:

1. Subtract the mean from each score $(X - \overline{X})$.
2. Then, square the resulting number to make it a positive number.

3. Then, add up these squared numbers to get a total score.
4. Then, take this total score and divide it by n − 1 (where n stands for the number of numbers that you have).
5. The final step is to take the square root of the number you found in step 4.

You will not be asked to compute the standard deviation using your calculator in this book, but you could see examples of how it is computed in any basic statistics book (e.g., Schuenemeyer and Drew 2011). Instead, we will use Excel to find the standard deviation of a set of scores. When we use Excel on the six numbers we gave in the description of the mean above, you will find that the *STDEV* of these numbers, S, is 1.47.

1.3   Standard Error of the Mean

The formula for the *standard error of the mean (s.e.*, which we will use $S_{\bar{X}}$ to symbolize) is:

$$\text{s.e.} = S_{\bar{X}} = \frac{S}{\sqrt{n}} \tag{1.3}$$

To find *s.e.*, all you need to do is to take the standard deviation, STDEV, and divide it by the square root of n, where n stands for the "number of numbers" that you have in your data set. In the example under the standard deviation description above, the *s.e.* = 0.60. (You can check this on your calculator.)

If you want to learn more about the standard deviation and the standard error of the mean, see McKillup and Dyar (2010) and Schuenemeyer and Drew (2011).

Now, let's learn how to use Excel to find the sample size, the mean, the standard deviation, and the standard error of the mean using the level of sulphur dioxide in rainfall measured in milligrams (mg) of sulphur per liter (L) of rainfall. (Note that one milligram (mg) equals one thousandth of one gram and is a metric measure of weight, while one liter is a metric unit of the volume of one kilogram of pure water under standard conditions.) Suppose that eight samples of rainfall were taken. The hypothetical data appear in Fig. 1.1.

Fig. 1.1 Worksheet Data
for Sulphur Dioxide Levels
(Practical Example)

	A	B	C	D
1				
2				
3		Sample	milligrams per liter (mg/L)	
4		1	0.4	
5		2	1.2	
6		3	0.8	
7		4	0.4	
8		5	1.3	
9		6	0.6	
10		7	0.7	
11		8	1.1	
12				
13				

1.4 Sample Size, Mean, Standard Deviation, and Standard Error of the Mean

> Objective: To find the sample size (n), mean, standard deviation (STDEV), and standard error of the mean (s.e.) for these data

Start your computer and click on the Excel 2019 icon to open a blank Excel spreadsheet.

Click on: Blank Workbook

Enter the data in this way:

B3: Sample
C3: milligrams per liter (mg/L)
B4   1

1.4.1 Using the Fill/Series/Columns Commands

> Objective: To add the sample numbers 2–8 in a column underneath Sample #1

Put pointer in B4

Home (top left of screen)

Important note: The "Paste" command should be on the top of your screen on the far left of the screen.

Important note: Notice the Excel commands at the top of your computer screen:
File → Home → Insert → Page Layout → Formulas → etc.
If these commands ever "disappear" when you are using Excel, you need to click on "Home" at the top left of your screen to make them reappear!

Fill (top right of screen: click on the down arrow; see Fig. 1.2)

Fig. 1.2 Home/Fill/Series commands

Series
Columns
Step value: 1
Stop value: 8 (see Fig. 1.3)

Fig. 1.3 Example of Dialog Box for Fill/Series/ Columns/Step Value/Stop Value commands

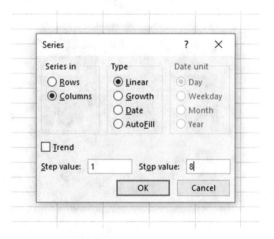

OK

The sample numbers should be identified as 1–8, with 8 in cell B11.

Now, enter the milligrams per liter in cells C4:C11. *(Note: Be sure to double-check your figures to make sure that they are correct or you will not get the correct answer!)*

Since your computer screen shows the information in a format that does not look professional, you need to learn how to "widen the column width" and how to "center the information" in a group of cells. Here is how you can do those two steps:

1.4.2   Changing the Width of a Column

Objective: To make a column width wider so that all of the information fits inside
              that column

If you look at your computer screen, you can see that Column C is not wide enough so that all of the information fits inside this column. To make Column C wider:

Click on the letter, C, at the top of your computer screen

Place your mouse pointer on your computer at the far right corner of C until you create a "cross sign" on that corner

Left-click on your mouse, hold it down, and move this corner to the right until it is "wide enough to fit all of the data"

Take your finger off your mouse to set the new column width (see Fig. 1.4)

Fig. 1.4 Example of How to Widen the Column Width

Then, click on any empty cell (i.e., any blank cell) to "deselect" column C so that it is no longer a darker color on your screen.

When you widen a column, you will make all of the cells in all of the rows of this column that same width.

Now, let's go through the steps to center the information in both Column B and Column C.

1.4.3   Centering Information in a Range of Cells

Objective: To center the information in a group of cells

In order to make the information in the cells look "more professional," you can center the information using the following steps:

Left-click your mouse pointer on B3 and drag it to the right and down to highlight
cells B3:C11 so that these cells appear in a darker color
Home

At the top of your computer screen, you will see a set of "lines" in which all of the lines are "centered" to the same width under "Alignment" (it is the second icon at the bottom left of the Alignment box; see Fig. 1.5)

Fig. 1.5 Example of How to Center Information Within Cells

Click on this icon to center the information in the selected cells (see Fig. 1.6).

Fig. 1.6 Final Result of
Centering Information in the
Cells

	A	B	C	D
1				
2				
3		Sample	milligrams per liter (mg/L)	
4		1	0.4	
5		2	1.2	
6		3	0.8	
7		4	0.4	
8		5	1.3	
9		6	0.6	
10		7	0.7	
11		8	1.1	
12				
13				

Since you will need to refer to the milligrams per liter in your formulas, it will be much easier to do this if you "name the range of data" with a name instead of having to remember the exact cells (C4:C11) in which these figures are located. Let's call that group of cells: Weight, but we could give them any name that you want to use.

1.4.4   Naming a Range of Cells

Objective: To name the range of data for the milligrams per liter with the name:
        Weight

Highlight cells C4:C11 by left-clicking your mouse pointer on C4 and dragging it
    down to C11
Formulas (top left of your screen)
Define Name (top center of your screen)
Weight (type this name in the top box; see Fig. 1.7)

Sample	milligrams per liter (mg/L)
1	0.4
2	1.2
3	0.8
4	0.4
5	1.3
6	0.6
7	0.7
8	1.1

New Name   ?   ×

Name: Weight

Scope: Workbook

Comment:

Refers to: =Sheet1!$C$4:$C$11

OK   Cancel

Fig. 1.7 Dialog box for "naming a range of cells" with the name: Weight

OK

Then, click on any cell of your spreadsheet that does not have any information in it (i.e., it is an "empty cell") to deselect cells C4:C11

Now, add the following terms to your spreadsheet:

E6:   n
E9:   Mean
E12:   STDEV
E15:   s.e. (see Fig. 1.8)

	A	B	C	D	E	F
1						
2						
3		Sample	milligrams per liter (mg/L)			
4		1	0.4			
5		2	1.2			
6		3	0.8		n	
7		4	0.4			
8		5	1.3			
9		6	0.6		Mean	
10		7	0.7			
11		8	1.1			
12					STDEV	
13						
14						
15					s.e.	
16						
17						

Fig. 1.8 Example of Entering the Sample Size, Mean, STDEV, and s.e. Labels

Note: Whenever you use a formula, you must add an equal sign (=) at the beginning of the name of the function so that Excel knows that you intend to use a formula.

1.4.5 Finding the Sample Size Using the =COUNT Function

> Objective: To find the sample size (n) for these data using the =COUNT function

F6:   =COUNT(Weight)

When you hit the Enter key, this command should insert the number 8 into cell F6 since there are eight samples of rainfall in your sample.

1.4.6 Finding the Mean Score Using the =AVERAGE Function

> Objective: To find the mean weight figure using the =AVERAGE function

F9:   =AVERAGE(Weight)

This command should insert the number 0.8125 into cell F9.

1.4.7 Finding the Standard Deviation Using the =STDEV Function

> Objective: To find the standard deviation (STDEV) using the =STDEV function

F12:   =STDEV(Weight)

This command should insert the number 0.352288 into cell F12.

1.4.8   Finding the Standard Error of the Mean

> Objective: To find the standard error of the mean using a formula for these eight
> data points

F15:   =F12/SQRT(8)

This command should insert the number 0.124553 into cell F15 (see Fig. 1.9).

	A	B	C	D	E	F	G
1							
2							
3		Sample	milligrams per liter (mg/L)				
4		1	0.4				
5		2	1.2				
6		3	0.8		n	8	
7		4	0.4				
8		5	1.3				
9		6	0.6		Mean	0.8125	
10		7	0.7				
11		8	1.1				
12					STDEV	0.352288	
13							
14							
15					s.e.	0.124553	
16							
17							

Fig. 1.9 Example of Using Excel Formulas for Sample Size, Mean, STDEV, and s.e.

*Important note: Throughout this book, be sure to double-check all of the figures in
your spreadsheet to make sure that they are in the correct cells, or
the formulas will not work correctly!*

1.4.8.1 Formatting Numbers in Number Format (Two Decimal Places)

Objective: To convert the mean, STDEV, and s.e. to two decimal places

Highlight cells F9:F15
Home (top left of screen)
Click on the down arrow to the right of "Number" at the top center of your screen
    Inside the dialog box, click on: Number
Keep the two decimal places already selected (see Fig. 1.10)

Fig. 1.10 Using the Number Format dialog box to convert Numbers to Fewer Decimal Places

OK (Notice that the cells F9:F15 are now all in two decimal places; see Fig. 1.11)

▲	A	B	C	D	E	F	G
1							
2							
3		Sample	milligrams per liter (mg/L)				
4		1	0.4				
5		2	1.2				
6		3	0.8		n	8	
7		4	0.4				
8		5	1.3				
9		6	0.6		Mean	0.81	
10		7	0.7				
11		8	1.1				
12					STDEV	0.35	
13							
14							
15					s.e.	0.12	
16							
17							

Fig. 1.11 Example of Converting Numbers to Two Decimal Places

Now, click on any "empty cell" on your spreadsheet to deselect cells F9:F15.

1.5 Saving a Spreadsheet

Objective: To save this spreadsheet with the name: sulphur3

In order to save your spreadsheet so that you can retrieve it sometime in the future, your first decision is to decide "where" you want to save it. That is your decision and you have several choices. If it is your own computer, you can save it onto your hard drive (you need to ask someone how to do that on your computer). Or, you can save it onto a "CD" or onto a "flash drive." You then need to complete these steps:

File
Save as

(select the place where you want to save the file by scrolling either down or up the bar on the left, and click on the place where you want to save the file; for example: This PC: Documents location)

File name: sulphur3 (enter this name to the right of File name; see Fig. 1.12)

Fig. 1.12 Dialog Box of Saving an Excel Workbook File as "sulphur3" in Documents location

Save

Important note: Be very careful to save your Excel file spreadsheet every few minutes so that you do not lose your information!

1.6 Printing a Spreadsheet

Objective: To print the spreadsheet

Use the following procedure when printing any spreadsheet:

File
Print
Print Active Sheets (see Fig. 1.13)

Fig. 1.13 Example of How to Print an Excel Worksheet Using the File/Print/Print Active Sheets Commands

Print (top of your screen)

The final spreadsheet is given in Fig. 1.14.

	A	B	C	D	E	F	G
1							
2							
3		Sample	milligrams per liter (mg/L)				
4		1	0.4				
5		2	1.2				
6		3	0.8		n	8	
7		4	0.4				
8		5	1.3				
9		6	0.6		Mean	0.81	
10		7	0.7				
11		8	1.1				
12					STDEV	0.35	
13							
14							
15					s.e.	0.12	
16							
17							

Fig. 1.14 Final Result of Printing an Excel Spreadsheet

Before you leave this chapter, let's practice changing the format of the figures on a spreadsheet with two examples: (1) using two decimal places for figures that are dollar amounts, and (2) using three decimal places for figures.

Save the final spreadsheet by: File/Save, then close your spreadsheet by: File/Close, and open a blank Excel spreadsheet by using File/New/Blank Workbook (on the top left of your screen).

1.7   Formatting Numbers in Currency Format (Two Decimal Places)

Objective: To change the format of figures to dollar format with two decimal places

A3:   Price
A4:   1.25
A5:   3.45
A6:   12.95

Highlight cells A4:A6 by left-clicking your mouse on A4 and dragging it down so that these three cells are highlighted in a darker color
Home
Number (top center of screen: click on the down arrow on the right; see Fig. 1.15)

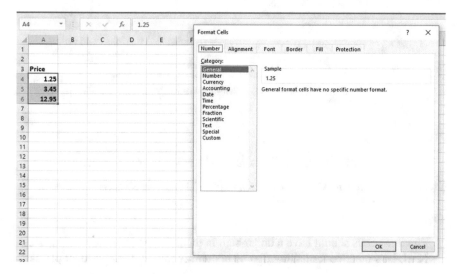

Fig. 1.15   Dialog Box for Number Format Choices

Category: Currency
Decimal places: 2 (then see Fig. 1.16)

Fig. 1.16 Dialog Box for Currency (two decimal places) Format for Numbers

OK

The three cells should have a dollar sign in them and be in two decimal places. Next, let's practice formatting figures in number format, three decimal places.

1.8   Formatting Numbers in Number Format (Three Decimal Places)

Objective: To format figures in number format, three decimal places

Home
Highlight cells A4:A6 on your computer screen
Number (click on the down arrow on the right)
Category: number
At the right of the box, change two decimal places to three decimal places by clicking on the "up arrow" once
OK

The three figures should now be in number format, each with three decimals.

Now, click on any blank cell to deselect cells A4:A6. Then, close this file by File/ Close/Don't Save (since there is no need to save this practice problem).

You can use these same commands to format a range of cells in percentage format (and many other formats) to whatever number of decimal places you want to specify.

1.9   End-of-Chapter Practice Problems

1. Suppose that you have been hired by the owner of a waterworks waste treatment plant to measure the chlorine residual each day in mg/L. You want to test your Excel skills, and so you have created the hypothetical data given in Fig. 1.17.

Fig. 1.17 Worksheet Data for Chap. 1: Practice Problem #1

CHLORINE RESIDUAL IN A WASTE TREATMENT PLANT
DATA (mg/L)
0.9
1.1
1.3
1.0
0.8
0.9
1.0
1.1
1.3
1.2
8.0
9.0
1.2
1.0
1.3
0.8
0.9

(a) Use Excel to the right of the table to find the sample size, mean, standard deviation, and standard error of the mean for these data. Label your answers and round off the mean, standard deviation, and standard error of the mean to two decimal places; use number format for these three figures.
(b) Print the result on a separate page.
(c) Save the file as: WATER4

2. Suppose that you have been hired as a research assistant and that you have been asked to determine the average micrograms of lead concentration per cubic meter ($\mu g/m^3$) for air samples taken near Route 101 near San Francisco in weekday afternoons between 4 p.m. and 7 p.m. The hypothetical data are given in Fig. 1.18.

Fig. 1.18 Worksheet Data for Chap. 1: Practice Problem #2

LEAD CONCENTRATION IN AIR SAMPLES TAKEN NEAR SAN FRANCISCO
Micrograms per cubic meter ($\mu g/m^3$)
3.1
10.1
6.7
8.9
5.6
6.4
4.8
10.2
9.8
8.4
7.5
9.4
8.5
4.8

(a) Use Excel to create a table of these data, and at the right of the table use Excel to find the sample size, mean, standard deviation, and standard error of the mean for these data. Label your answers and round off the mean, standard deviation, and standard error of the mean to two decimal places using number format.

(b) Print the result on a separate page.

(c) Save the file as: air3

3. Measurements taken on environmental variables vary with each measurement attempt. Suppose that you wanted to establish measurements for tetrachlorobenzene (TcCB) in an uncontaminated site so that you could use the measurements taken from different locations in this site as a reference for testing future sites in terms of their possible contamination. The hypothetical data measured in parts per billion (ppb) from a site are given in Fig. 1.19:

Fig. 1.19 Worksheet Data for Chap. 1: Practice Problem #3

MEASUREMENTS OF TcCB IN AN UNCONTAMINATED SITE

parts per billion (ppb)
0.25
0.26
0.28
0.24
1.11
1.31
1.21
0.72
0.84
0.28
0.53
0.86

(a) Use Excel to create a table for these data, and at the right of the table, use Excel to find the sample size, mean, standard deviation, and standard error of the mean for these data. Label your answers and round off the mean, standard deviation, and standard error of the mean to three decimal places using number format.

(b) Print the result on a separate page.

(c) Save the file as: SITE3

References

McKillup S., Dyar M. Geostatistics Explained: an introductory guide for earth scientists. Cambridge: Cambridge University Press; 2010.

Schuenemeyer J, Drew L. Statistics for Earth and Environmental Scientists. Hoboken: John Wiley & Sons; 2011.

Chapter 2
Random Number Generator

Salt marshes are coastal wetlands found on protected shorelines along the eastern seaboard of the USA where freshwater mixes with seawater. When ocean tides flood salt marshes, the plants living there must cope with the salt water. The "salinity" (i.e., the salt content of the water) depends on how close the marsh is to the ocean. Suppose that a biogeographer is studying the effects of salinity on vegetation in a salt marsh in Maine and that she has mapped the salt marsh into 32 separate geographic areas. Suppose, further, that she has asked you to take a random sample of 5 of these 32 areas within the salt marsh so that she can measure the percent of salinity level in each of these areas. Using your Excel skills to take this random sample, you will need to define a "sampling frame."

A sampling frame is a list of objects, events, or people from which you want to select a random sample. In this case, it is the group of 32 areas of the salt marsh. The frame starts with the identification code (ID) of the number 1 that is assigned to the first area in the group of 32 areas. The second area has a code number of 2, the third a code number of 3, and so forth until the last area has a code number of 32.

Since the salt marsh has 32 areas, your sampling frame would go from 1 to 32 with each area having a unique ID number.

We will first create the frame numbers as follows in a new Excel worksheet:

2.1 Creating Frame Numbers for Generating Random Numbers

Objective: To create the frame numbers for generating random numbers

A3: FRAME NO.
A4: 1

© Springer Nature Switzerland AG 2021
T. J. Quirk et al., *Excel 2019 for Environmental Sciences Statistics*,
Excel for Statistics, https://doi.org/10.1007/978-3-030-66277-6_2

Now, create the frame numbers in column A with the Home/Fill commands that were explained in the first chapter of this book (see Sect. 1.4.1) so that the frame numbers go from 1 to 32, with the number 32 in cell A35. If you need to be reminded about how to do that, here are the steps:

Click on cell A4 to select this cell
Home
Fill (then click on the "down arrow" next to this command)
(see Fig. 2.1)

Fig. 2.1 Dialog Box for
Fill/Series Commands

Series
Columns
Step value: 1
Stop value: 32 (see Fig. 2.2)

Fig. 2.2 Dialog Box for
Fill/Series/Columns/Step
value/Stop value
Commands

OK

Then, save this file as: Random29. You should obtain the result in Fig. 2.3.

Fig. 2.3 Frame Numbers
from 1 to 32

FRAME NO.
1
2
3
4
5
6
7
8
9
10
11
12
13
14
15
16
17
18
19
20
21
22
23
24
25
26
27
28
29
30
31
32

Now, create a column next to these frame numbers in this manner:

B3:   DUPLICATE FRAME NO.
B4:   1

Next, use the Home/Fill command again, so that the 32 frame numbers begin in cell B4 and end in cell B35. Be sure to widen the columns A and B so that all of the information in these columns fits inside the column width. Then, center the information inside both Column A and Column B on your spreadsheet. You should obtain the information given in Fig. 2.4.

Fig. 2.4 Duplicate Frame
Numbers from 1 to 32

FRAME NO.	DUPLICATE FRAME NO.
1	1
2	2
3	3
4	4
5	5
6	6
7	7
8	8
9	9
10	10
11	11
12	12
13	13
14	14
15	15
16	16
17	17
18	18
19	19
20	20
21	21
22	22
23	23
24	24
25	25
26	26
27	27
28	28
29	29
30	30
31	31
32	32

Save this file as: Random30

You are probably wondering why you created the same information in both Column A and Column B of your spreadsheet. This is to make sure that before you sort the frame numbers that you have exactly 32 of them when you finish sorting them into a random sequence of 32 numbers.

Now, let's add a random number to each of the duplicate frame numbers as follows:

2.2 Creating Random Numbers in an Excel Worksheet

C3:   RANDOM NO. (then widen columns A, B, C so that their labels fit inside the columns; then center the information in A3:C35)

C4:   =RAND()

Next, hit the Enter key to add a random number to cell C4.

Note that you need *both* an open parenthesis *and* a closed parenthesis after *=RAND()*. The RAND command "looks to the left of the cell with the RAND() COMMAND in it" and assigns a random number to that cell.

Now, put the pointer using your mouse in cell C4 and then move the pointer to the bottom right corner of that cell until you see a "plus sign" in that cell. Then, click and drag the pointer down to cell C35 to add a random number to all 32 ID frame numbers (see Fig. 2.5).

Fig. 2.5 Example of Random Numbers Assigned to the Duplicate Frame Numbers

FRAME NO.	DUPLICATE FRAME NO.	RANDOM NO.
1	1	0.178997426
2	2	0.269196787
3	3	0.48649709
4	4	0.882904516
5	5	0.015953504
6	6	0.099651545
7	7	0.42850057
8	8	0.381659988
9	9	0.431296832
10	10	0.476642453
11	11	0.268603728
12	12	0.871330234
13	13	0.775421903
14	14	0.908450998
15	15	0.138749452
16	16	0.159535582
17	17	0.672417279
18	18	0.956231064
19	19	0.486746795
20	20	0.83596565
21	21	0.688574546
22	22	0.467838617
23	23	0.695493167
24	24	0.226521237
25	25	0.335451708
26	26	0.209245145
27	27	0.631291464
28	28	0.210229448
29	29	0.553196562
30	30	0.494647331
31	31	0.986702143
32	32	0.178067956

Then, click on any empty cell to deselect C4:C35 to remove the dark color highlighting these cells.

Save this file as: Random31

Now, let's sort these duplicate frame numbers into a random sequence:

2.3   Sorting Frame Numbers into a Random Sequence

Objective: To sort the duplicate frame numbers into a random sequence

Highlight cells B3:C35 (include the labels at the top of columns B and C)
Data (top of screen)
Sort (click on this word at the top center of your screen; see Fig. 2.6)

Fig. 2.6 Dialog Box for
Data/Sort Commands

Sort by: RANDOM NO. (click on the down arrow)
Smallest to Largest (see Fig. 2.7)

Fig. 2.7   Dialog Box for Data/Sort/RANDOM NO./Smallest to Largest Commands

OK

Click on any empty cell to deselect B3:C35.

Save this file as: Random32

Print this file now.

These steps will produce Fig. 2.8 with the DUPLICATE FRAME NUMBERS sorted into a random order:

Important note: Because Excel randomly assigns these random numbers, your Excel commands will produce a different sequence of random numbers from everyone else who reads this book!

Fig. 2.8 Duplicate Frame Numbers Sorted by Random Number

FRAME NO.	DUPLICATE FRAME NO.	RANDOM NO.
1	5	0.063981403
2	6	0.977468743
3	15	0.225170263
4	16	0.765734052
5	32	0.274680922
6	1	0.594468001
7	26	0.511966171
8	28	0.625577233
9	24	0.906310053
10	11	0.488640116
11	2	0.020129977
12	25	0.723003676
13	8	0.975227547
14	7	0.469582962
15	9	0.14889954
16	22	0.955629903
17	10	0.897398234
18	3	0.314860892
19	19	0.442019486
20	30	0.078566335
21	29	0.172474705
22	27	0.104689528
23	17	0.406630369
24	21	0.961398315
25	23	0.094222677
26	13	0.323429051
27	20	0.470615753
28	12	0.978014724
29	4	0.618082813
30	14	0.727776384
31	18	0.919475329
32	31	0.324497007

Because your objective at the beginning of this chapter was to select randomly 5 of the 32 areas of the salt marsh, you now can do that by selecting the *first five ID numbers* in DUPLICATE FRAME NO. column after the sort.

Although your first five random numbers will be different from those we have selected in the random sort that we did in this chapter, we would select these five IDs of areas using Fig. 2.9.

Fig. 2.9 First Five Areas Selected Randomly

FRAME NO.	DUPLICATE FRAME NO.	RANDOM NO.
1	5	0.063981403
2	6	0.977468743
3	15	0.225170263
4	16	0.765734052
5	32	0.274680922
6	1	0.594468001
7	26	0.511966171
8	28	0.625577233
9	24	0.906310053
10	11	0.488640116
11	2	0.020129977
12	25	0.723003676
13	8	0.975227547
14	7	0.469582962
15	9	0.14889954
16	22	0.955629903
17	10	0.897398234
18	3	0.314860892
19	19	0.442019486
20	30	0.078566335
21	29	0.172474705
22	27	0.104689528
23	17	0.406630369
24	21	0.961398315
25	23	0.094222677
26	13	0.323429051
27	20	0.470615753
28	12	0.978014724
29	4	0.618082813
30	14	0.727776384
31	18	0.919475329
32	31	0.324497007

5, 6, 15, 16, 32
Save this file as: Random33

Remember, your five ID numbers selected after your random sort will be different from the five ID numbers in Fig. 2.9 because Excel assigns a different random number *each time the =RAND() command is given.*

Before we leave this chapter, you need to learn how to print a file so that all of the information on that file fits onto a single page without "dribbling over" onto a second or third page.

2.4 Printing an Excel File So That All of the Information Fits onto One Page

Objective: To print a file so that all of the information fits onto one page

Note that the three practice problems at the end of this chapter require you to sort random numbers when the files contain 42 water samples, 86 employees, and 75 toxic waste sites, respectively. These files will be "too big" to fit onto one page when you print them unless you format these files so that they fit onto a single page when you print them.

Let's create a situation where the file does not fit onto one printed page unless you format it first to do that.

Go back to the file you just created, Random 33, and enter the name: *Jennifer* into cell: A52.

If you printed this file now, the name, *Jennifer*, would be printed onto a second page because it "dribbles over" outside of the page rage for this file in its current format.

So, you would need to change the page format so that all of the information, including the name, Jennifer, fits onto just one page when you print this file by using the following steps:

Page Layout (top left of the computer screen)
(Notice the "Scale to Fit" section in the center of your screen; see Fig. 2.10)

Fig. 2.10 Dialog Box for Page Layout/Scale to Fit Commands

Hit the down arrow to the right of 100% *once* to reduce the size of the page to 95%.

Now, note that the name, Jennifer, is still on a second page on your screen because her name is below the horizontal dotted line on your screen in Fig. 2.11 (the dotted lines tell you outline dimensions of the file if you printed it now).

File	Home	Insert	Page Layout		Formulas	Data	Review	View	Help	Tell me what

Themes · Colors ▾ / A Fonts ▾ / Effects ▾ | Margins ▾ | Orientation ▾ | Size ▾ | Print Area ▾ | Breaks ▾ | Background | Print Titles | Width: Automatic ▾ | Height: Automatic ▾ | Scale: 95%

	Themes			Page Setup			Scale to Fit	
29	26	14	0.144288401					
30	27	9	0.712622785					
31	28	26	0.908952646					
32	29	20	0.720731489					
33	30	2	0.352807163					
34	31	5	0.378405493					
35	32	15	0.368357556					
36								
37								
38								
39								
40								
41								
42								
43								
44								
45								
46								
47								
48								
49								
50								
51								
52	*Jennifer*							
53								
54								
55								

Fig. 2.11 Example of Scale Reduced to 95% with "Jennifer" to be Printed on a Second Page

So, you need to repeat the "scale change steps" by hitting the down arrow on the right once more to reduce the size of the worksheet to 90% of its normal size.

Notice that the "dotted lines" on your computer screen in Fig. 2.12 are now below Jennifer's name to indicate that all of the information, including her name, is now formatted to fit onto just one page when you print this file.

Fig. 2.12 Example of Scale Reduced to 90% with "Jennifer" to be printed on the first page (note the dotted line below Jennifer on your screen)

Save the file as: Random34

Print the file. Does it all fit onto one page? It should (see Fig. 2.13).

Fig. 2.13 Final
Spreadsheet of 90% Scale
to Fit

FRAME NO.	DUPLICATE FRAME NO.	RANDOM NO.
1	5	0.747176905
2	6	0.038774393
3	15	0.091368861
4	16	0.63147137
5	32	0.190734495
6	1	0.411943765
7	26	0.138033007
8	28	0.927874602
9	24	0.058336576
10	11	0.043243606
11	2	0.729011126
12	25	0.204119693
13	8	0.456656709
14	7	0.232589896
15	9	0.09096704
16	22	0.935399501
17	10	0.201267198
18	3	0.52638312
19	19	0.53734605
20	30	0.969840616
21	29	0.475657455
22	27	0.558049277
23	17	0.488444809
24	21	0.717097206
25	23	0.86192944
26	13	0.875595013
27	20	0.536748908
28	12	0.331784725
29	4	0.642847666
30	14	0.575767804
31	18	0.939789757
32	31	0.776050794
Jennifer		

2.5   End-of-Chapter Practice Problems

1. Suppose that you were hired to test the fluoride levels in drinking water in Jefferson County, Colorado. Historically, there are a total of 42 water sample collection sites. Because of budget constraints, you need to choose a random sample of 12 of these 42 water sample collection sites.

   (a) Set up a spreadsheet of frame numbers for these water samples with the heading: FRAME NUMBERS
   (b) Then, create a separate column to the right of these frame numbers which duplicates these frame numbers with the title: Duplicate frame numbers
   (c) Then, create a separate column to the right of these duplicate frame numbers called RAND NO. and use the =RAND() function to assign random numbers to all of the frame numbers in the duplicate frame numbers column and change this column format so that three decimal places appear for each random number.
   (d) Sort the *duplicate frame numbers and random numbers* into a random order.
   (e) Print the result so that the spreadsheet fits onto one page.
   (f) Circle on your printout the I.D. number of the first 12 water sample locations that you would use in your test.
   (g) Save the file as: RAND13

   Important note: Note that everyone who does this problem will generate a different random order of water sample sites ID numbers since Excel assigns a different random number each time the RAND() command is used. For this reason, the answer to this problem given in this Excel Guide will have a completely different sequence of random numbers from the random sequence that you generate. This is normal and what is to be expected.

2. Suppose that you are the Human Resource Director (HR) at a large Environmental Science company and that you want to conduct a phone interview with 25 of the 86 employees who elected to be included in the company's new Dental Care Plan which was instituted a year ago to obtain their feedback about how well the plan has been working for them.

   (a) Set up a spreadsheet of frame numbers for these employees with the heading: FRAME NUMBERS
   (b) Then, create a separate column to the right of these frame numbers which duplicates these frame numbers with the title: Duplicate frame numbers
   (c) Then, create a separate column to the right of these duplicate frame numbers entitled "Random number" and use the =RAND() function to assign random numbers to all of the frame numbers in the duplicate frame numbers column. Then, change this column format so that three decimal places appear for each random number.
   (d) Sort the duplicate frame numbers and random numbers into a random order.

(e) Print the result so that the spreadsheet fits onto one page.

(f) Circle on your printout the I.D. number of the first 25 employees that the HR Director should select for her study.

(g) Save the file as: RAND14

3. Suppose that a chemical field researcher wants to take a random sample of 20 of 75 toxic waste sites that have been mapped surrounding a commercial house paint plant that has been closed and abandoned. The researcher wants to test the amount of lead in the soil around this plant as part of a field research study.

(a) Set up a spreadsheet of frame numbers for these sites with the heading: FRAME NUMBERS

(b) Then, create a separate column to the right of these frame numbers which duplicates these frame numbers with the title: Duplicate frame numbers

(c) Then, create a separate column to the right of these duplicate frame numbers entitled "Random number" and use the =RAND() function to assign random numbers to all of the frame numbers in the duplicate frame numbers column. Then, change this column format so that three decimal places appear for each random number.

(d) Sort the duplicate frame numbers and random numbers into a random order.

(e) Print the result so that the spreadsheet fits onto one page.

(f) Circle on your printout the I.D. number of the first 20 sites that the field chemist should select for her study.

(g) Save the file as: RAND5

Chapter 3
Confidence Interval About the Mean Using the TINV Function and Hypothesis Testing

This chapter focuses on two ideas: (1) finding the 95% confidence interval about the mean and (2) hypothesis testing.

Let's talk about the confidence interval first.

3.1 Confidence Interval About the Mean

In statistics, we are always interested in *estimating the population mean*. How do we do that?

3.1.1 How to Estimate the Population Mean

Objective: To estimate the population mean, μ

Remember that the population mean is the average of all of the people in the target population. For example, if we were interested in how well adults ages 25–44 liked a new flavor of Ben and Jerry's ice cream, we could never ask this question of all of the people in the USA who were in that age group. Such a research study would take way too much time to complete and the cost of doing that study would be prohibitive.

So, instead of testing *everyone* in the population, we take a sample of people in the population and use the results of this sample to estimate the mean of the entire population. This saves both time and money. When we use the results of a sample to estimate the population mean, this is called "*inferential statistics*" because we are inferring the population mean from the sample mean.

© Springer Nature Switzerland AG 2021

T. J. Quirk et al., *Excel 2019 for Environmental Sciences Statistics*, Excel for Statistics, https://doi.org/10.1007/978-3-030-66277-6_3

When we study a sample of people in science research, we know the size of our sample (n), the mean of our sample ($\overline{X}$), and the standard deviation of our sample (STDEV). We use these figures to estimate the population mean with a test called the "confidence interval about the mean."

3.1.2   Estimating the Lower Limit and the Upper Limit of the 95% Confidence Interval About the Mean

The theoretical background of this test is beyond the scope of this book, and you can learn more about this test from studying any good statistics textbook (e.g., Levine (2011) or Bremer and Doerge (2010)), but the basic ideas are as follows.

We assume that the population mean is somewhere in an interval which has a "lower limit" and an "upper limit" to it. We also assume in this book that we want to be "95% confident" that the population mean is inside this interval somewhere. So, we intend to make the following type of statement:

"We are 95% confident that the population mean in miles per gallon (mpg) for the Chevy Impala automobile is between 26.92 miles per gallon and 29.42 miles per gallon."

If we want to create a billboard emphasizing the perceived lower environmental impact of the Chevy Impala by claiming that this car gets 28 miles per gallon (mpg), we can do this because 28 is *inside the 95% confidence interval* in our research study in the above example. We do not know exactly what the population mean is, only that it is somewhere between 26.92 mpg and 29.42 mpg, and 28 is inside this interval.

But we are only 95% confident that the population mean is inside this interval, and 5% of the time we will be wrong in assuming that the population mean is 28 mpg.

But, for our purposes in science research, we are happy to be 95% confident that our assumption is accurate. We should also point out that 95% is an arbitrary level of confidence for our results. We could choose to be 80% confident, or 90% confident, or even 99% confident in our results if we wanted to do that. But, in this book, *we will always assume that we want to be 95% confident of our results.* That way, you will not have to guess on how confident you want to be in any of the problems in this book. We will always want to be 95% confident of our results in this book.

So how do we find the 95% confidence interval about the mean for our data?

In words, we will find this interval this way:

"Take the sample mean ($\overline{X}$), *and add to it* 1.96 times the standard error of the mean (s.e.) to get the upper limit of the confidence interval. Then, take the sample mean, *and subtract from it* 1.96 times the standard error of the mean to get the lower limit of the confidence interval."

You will remember (See Sect. 1.3) that the standard error of the mean (s.e.) is found by dividing the standard deviation of our sample (STDEV) by the square root of our sample size, n.

In mathematical terms, the formula for the 95% confidence interval about the mean is:

$$\overline{X} \pm 1.96 \text{ s.e.} \qquad\qquad (3.1)$$

Note that the " $\pm$ *sign*" stands for "plus or minus," and this means that you first add 1.96 times the s.e. to the mean to get the upper limit of the confidence interval, and then subtract 1.96 times the s.e. from the mean to get the lower limit of the confidence interval. Also, the symbol 1.96 s.e. means that you multiply 1.96 times the standard error of the mean to get this part of the formula for the confidence interval.

Note: We will explain shortly where the number 1.96 came from.

Let's try a simple example to illustrate this formula.

3.1.3 Estimating the Confidence Interval for the Chevy Impala in Miles Per Gallon

Let's suppose that you have been asked to be a part of a larger study looking at the carbon footprint of Chevy Impala drivers. You are interested in the average miles per gallon (mpg) of a Chevy Impala. You asked owners of the Chevy Impala to keep track of their mileage and the number of gallons used for two tanks of gas. Let's suppose that 49 owners did this, and that they average 27.83 miles per gallon (mpg) with a standard deviation of 3.01 mpg. The standard error (s.e.) would be 3.01 divided by the square root of 49 (i.e., 7) which gives a s.e. equal to 0.43.

The 95% confidence interval for these data would be:

$$27.83 \pm 1.96 \, (0.43)$$

The *upper limit of this confidence interval* uses the plus sign of the $\pm$ sign in the formula. Therefore, the upper limit would be:

$$27.83 + 1.96 \, (0.43) = 27.83 + 0.84 = 28.67 \text{ mpg}$$

Similarly, *the lower limit of this confidence interval* uses the minus sign of the $\pm$ sign in the formula. Therefore, the lower limit would be:

$$27.83 - 1.96 \, (0.43) = 27.83 - 0.84 = 26.99 \text{ mpg}$$

The result of our part of the ongoing research study would, therefore, be the following:

"We are 95% confident that the population mean for the Chevy Impala is somewhere between 26.99 mpg and 28.67 mpg."

Based upon the 28 mpg of the Chevy Impala, we could create a billboard emphasizing the higher miles per gallon and highlight a perceived lower environmental impact. Our data supports this claim because the 28 mpg is inside of this 95% confidence interval for the population mean.

You are probably asking yourself: "Where did that 1.96 in the formula come from?"

3.1.4 Where Did the Number "1.96" Come From?

A detailed mathematical answer to that question is beyond the scope of this book, but here is the basic idea.

We make an assumption that the data in the population are "normally distributed" in the sense that the population data would take the shape of a "normal curve" if we could test all of the people in the population. The normal curve looks like the outline of the Liberty Bell that sits in front of Independence Hall in Philadelphia, Pennsylvania. The normal curve is "symmetric" in the sense that if we cut it down the middle, and folded it over to one side, the half that we folded over would fit perfectly onto the half on the other side (see Webster and Oliver 2007).

A discussion of integral calculus is beyond the scope of this book, but essentially we want to find the lower limit and the upper limit of the population data in the normal curve so that 95% of the area under this curve is between these two limits. *If we have more than 40 people in our research study*, the value of these limits is plus or minus 1.96 times the standard error of the mean (s.e.) of our sample. The number 1.96 times the s.e. of our sample gives us the upper limit and the lower limit of our confidence interval. If you want to learn more about this idea, you can consult a good statistics book (e.g., Schuenemeyer and Drew 2011).

The number 1.96 would change if we wanted to be confident of our results at a different level from 95% as long as we have more than 40 people in our research study.

For example:

1. If we wanted to be 80% confident of our results, this number would be 1.282.
2. If we wanted to be 90% confident of our results, this number would be 1.645.
3. If we wanted to be 99% confident of our results, this number would be 2.576.

But since we always want to be 95% confident of our results in this book, we will always use 1.96 in this book whenever we have more than 40 people in our research study.

By now, you are probably asking yourself: "Is this number in the confidence interval about the mean always 1.96?" The answer is: "No!" and we will explain why this is true now.

3.1.5   Finding the Value for t in the Confidence Interval Formula

---
Objective: To find the value for t in the confidence interval formula

---

The correct formula for the confidence interval about the mean for different sample sizes is the following:

$$\overline{X} \pm t \text{ s.e.} \tag{3.2}$$

To use this formula, you find the sample mean, $\overline{X}$, *and add to it the value of t times the s.e. to get the upper limit* of this 95% confidence interval. Also, you take the sample mean, $\overline{X}$, and *subtract from it the value of t times the s.e. to get the lower limit* of this 95% confidence interval. And, you find the value of t in the table given in Appendix E of this book in the following way:

Objective: To find the value of t in the t-table in Appendix E

Before we get into an explanation of what is meant by "the value of t," let's give you practice in finding the value of t by using the t-table in Appendix E.

Keep your finger on Appendix E as we explain how you need to "read" that table.

Since the test in this chapter is called the "confidence interval about the mean test," you will use the first column on the left in Appendix E to find the critical value of t for your research study (note that this column is headed: "sample size n").

To find the value of *t*, you go down this first column until you find the sample size in your research study, and then you go to the right and read the value of *t* for that sample size in the "critical t column" of the table (note that this column is the column that you would use for the 95% confidence interval about the mean).

For example, if you have 14 people in your research study, the value of t is 2.160.

If you have 26 people in your research study, the value of t is 2.060.

If you have more than 40 people in your research study, the value of t is always 1.96.

Note that the "critical t column" in Appendix E represents the value of t that you need to use to obtain to be 95% confident of your results as "significant" results.

Throughout this book, we are assuming that you want to be 95% confident in the results of your statistical tests. Therefore, the value for t in the t-table in Appendix E tells you which value you should use for *t* when you use the formula for the 95% confidence interval about the mean.

Now that you know how to find the value of t in the formula for the confidence interval about the mean, let's explore how you find this confidence interval using Excel.

3.1.6   Using Excel's TINV Function to Find the Confidence Interval About the Mean

> Objective: To use the TINV function in Excel to find the confidence interval about the mean

When you use Excel, the formulas for finding the confidence interval are:

$$Lower\ limit = \overline{X} - TINV(1 - 0.95, n - 1) * s.e. \quad \text{(no spaces between these symbols)}$$
$$(3.3)$$

$$Upper\ limit = \overline{X} + TINV(1 - 0.95, n - 1) * s.e. \quad \text{(no spaces between these symbols)}$$
$$(3.4)$$

Note that the "* *symbol*" in this formula tells Excel to use the multiplication step in the formula, and it stands for "times" in the way we talk about multiplication.

You will recall from Chap. 1 that n stands for the sample size, and so $n - 1$ stands for the sample size minus one.

You will also recall from Chap. 1 that the standard error of the mean, s.e., equals the STDEV divided by the square root of the sample size, n (See Sect. 1.3).

Let's try a sample problem using Excel to find the 95% confidence interval about the mean for a problem.

Let's suppose that General Motors wanted to claim that its Chevy Impala achieves 28 miles per gallon (mpg). Let's call 28 mpg the "reference value" for this car.

Suppose that you work for Ford Motor Co. and that you want to check this claim to see if it holds up based on some research evidence. You decide to collect some data and to use a two-side 95% confidence interval about the mean to test your results:

3.1.7   Using Excel to Find the 95% Confidence Interval for a Car's mpg Claim

> Objective: To analyze the data using a two-side 95% confidence interval about the mean

You select a sample of new car owners for this car and they agree to keep track of their mileage for two tanks of gas and to record the average miles per gallon they achieve on these two tanks of gas. Your research study produces the results given in Fig. 3.1:

Chevy Impala					
Miles per gallon					
30.9					
24.5					
31.2					
28.7					
35.1					
29.0					
28.8					
23.1					
31.0					
30.2					
28.4					
29.3					
24.2					
27.0					
26.7					
31.0					
23.5					
29.4					
26.3					
27.5					
28.2					
28.4					
29.1					
21.9					
30.9					

Fig. 3.1 Worksheet Data for Chevy Impala (Practical Example)

Create a spreadsheet with these data and use Excel to find the sample size (n), the mean, the standard deviation (STDEV), and the standard error of the mean (s.e.) for these data using the following cell references.

A3:   Chevy Impala
A5:   Miles per gallon
A6:   30.9

Enter the other mpg data in cells A7:A30

Now, highlight cells A6:A30 and format these numbers in number format (one decimal place). Center these numbers in Column A. Then, widen columns A and B by making both of them twice as wide as the original width of column A. Then, widen column C so that it is three times as wide as the original width of column A so that your table looks more professional.

C7: n
C10: Mean
C13: STDEV
C16: s.e.
C19: 95% confidence interval
D21: Lower limit:
D23: Upper limit: (see Fig. 3.2)

Chevy Impala			
Miles per gallon			
30.9			
24.5	n		
31.2			
28.7			
35.1	Mean		
29.0			
28.8			
23.1	STDEV		
31.0			
30.2			
28.4	s.e		
29.3			
24.2			
27.0	95% confidence interval		
26.7			
31.0			Lower limit:
23.5			
29.4			Upper Limit:
26.3			
27.5			
28.2			
28.4			
29.1			
21.9			
30.9			

Fig. 3.2 Example of Chevy Impala Format for the Confidence Interval About the Mean Labels

B26: Draw a picture below this confidence interval
B28: 26.92
B29: lower (right-align this word)
B30: limit (right-align this word)
C28: '-------- 28 -------28.17 ---------- (note that you need to begin cell C28 with a
   single quotation mark (') to tell Excel that this is a *label*, and not a number)

D28:   ' ---------- (note the single quotation mark)
E28:   '29.42 (note the single quotation mark)
C29:   ref.   Mean
C30:   value
E29:   upper
E30:   limit
B33:   Conclusion:

Now, align the labels underneath the picture of the confidence interval so that they look like Fig. 3.3.

Chevy Impala				
Miles per gallon				
30.9				
24.5		n		
31.2				
28.7				
35.1		**Mean**		
29.0				
28.8				
23.1		**STDEV**		
31.0				
30.2				
28.4		s.e		
29.3				
24.2				
27.0		**95% confidence interval**		
26.7				
31.0			**Lower limit:**	
23.5				
29.4			**Upper Limit:**	
26.3				
27.5				
28.2	**Draw a picture below this confidence interval**			
28.4				
29.1	26.92 ---------- 28 ---------- 28.17 ---------------- 29.42			
21.9	lower ref. Mean upper			
30.9	limit value limit			
	Conclusion:			

Fig. 3.3 Example of Drawing a Picture of a Confidence Interval About the Mean Result

Next, name the range of data from A6:A30 as: miles

D7:   Use Excel to find the sample size
D10: Use Excel to find the mean

D13:   Use Excel to find the STDEV
D16:   Use Excel to find the s.e.

Now, you need to find the lower limit and the upper limit of the 95% confidence interval for this study.

We will use Excel's TINV function to do this. We will assume that you want to be 95% confident of your results.

$$F21 : = D10 - TINV(1 - .95, 24) * D16 \quad \text{(no spaces between symbols)}$$

Note that this TINV formula uses 24 since 24 is one less than the sample size of 25 (i.e., 24 is n − 1). Note that D10 is the mean, while D16 is the standard error of the mean. The above formula gives the *lower limit of the confidence interval, 26.92.*

$$F23 : = D10 + TINV(1 - .95, 24) * D16 \quad \text{(no spaces between symbols)}$$

The above formula gives the *upper limit of the confidence interval, 29.42.*

Now, use number format (two decimal places) in your Excel spreadsheet for the mean, standard deviation, standard error of the mean, and for both the lower limit and the upper limit of your confidence interval. If you printed this spreadsheet now, the lower limit of the confidence interval (26.92) and the upper limit of the confidence interval (29.42) would "dribble over" onto a second printed page because the information on the spreadsheet is too large to fit onto one page in its present format.

So, you need to use Excel's "Scale to Fit" commands that we discussed in Chap. 2 (see Sect. 2.4) to reduce the size of the spreadsheet to 95% of its current size using the Page Layout/Scale to Fit function. Do that now and notice that the dotted line to the right of 26.92 and 29.42 indicates that these numbers would now fit onto one page when the spreadsheet is printed out (see Fig. 3.4).

Chevy Impala				
Miles per gallon				
30.9				
24.5		n		25
31.2				
28.7				
35.1		**Mean**		28.17
29.0				
28.8				
23.1		**STDEV**		3.03
31.0				
30.2				
28.4		s.e		0.61
29.3				
24.2				
27.0		**95% confidence interval**		
26.7				
31.0			**Lower limit:**	26.92
23.5				
29.4			**Upper Limit:**	29.42
26.3				
27.5				
28.2		**Draw a picture below this confidence interval**		
28.4				
29.1		26.92 ---------- 28 ---------- 28.17 -------------- 29.42		
21.9		lower ref. Mean upper		
30.9		limit value limit		
	Conclusion:			

Fig. 3.4 Result of Using the TINV Function to Find the Confidence Interval About the Mean

Note that you have drawn a picture of the 95% confidence interval beneath cell B26, including the lower limit, the upper limit, the mean, and the reference value of 28 mpg given in the claim that the company wants to make about the car's miles per gallon performance.

Now, let's write the conclusion to your research study on your spreadsheet:

C33: Since the reference value of 28 is inside
C34: the confidence interval, we accept that
C35: the Chevy Impala does get 28 mpg.

Important note: You are probably wondering why we wrote the conclusion on three separate lines of the spreadsheet instead of writing it on one long line. This is because if you wrote it on one line, two things would happen that you would not like: (1) If you printed the conclusion by

reducing the size of the layout of the page so that the entire spreadsheet would fit onto one page, the print font size for the entire spreadsheet would be so small that you could not read it without a magnifying glass, and (2) If you printed the spreadsheet without reducing the page size layout, it would "dribble over" part of the conclusion to a separate page all by itself, and your spreadsheet would not look professional.

Your research study accepted the claim that the Chevy Impala did get 28 miles per gallon. The average miles per gallon in your study was 28.17. (See Fig. 3.5)

Save your resulting spreadsheet as: **CHEVY7**

Chevy Impala			
Miles per gallon			
30.9			
24.5	n		25
31.2			
28.7			
35.1	Mean		28.17
29.0			
28.8			
23.1	STDEV		3.03
31.0			
30.2			
28.4	s.e		0.61
29.3			
24.2			
27.0	95% confidence interval		
26.7			
31.0		Lower limit:	26.92
23.5			
29.4		Upper Limit:	29.42
26.3			
27.5			
28.2	Draw a picture below this confidence interval		
28.4			
29.1	26.92 ---------- 28 ---------- 28.17 --------------- 29.42		
21.9	lower ref. Mean upper		
30.9	limit value limit		
	Conclusion:	Since the reference value of 28 is inside the confidence interval, we accept that the Chevy Impala does get 28 mpg.	

Fig. 3.5 Final Spreadsheet for the Chevy Impala Confidence Interval About the Mean

3.2 Hypothesis Testing

One of the important activities of research scientists is that they attempt to "check" their assumptions about the world by testing these assumptions in the form of hypotheses.

A typical hypothesis is in the form: *"If x, then y."*

Some examples would be:

1. "If we use this new method fertilizing the soil, the corn yield of the plot will increase by 3%."
2. "If we use this new method of teaching science to ninth graders, then our science achievement scores will go up by 5%."
3. "If we change the format for teaching Introductory Biology to our undergraduates, then their final exam scores will increase by 8%."

A hypothesis, then, to a research scientist is a "guess" about what we think is true in the real world. We can test these guesses using statistical formulas to see if our predictions come true in the real world.

So, in order to perform these statistical tests, we must first state our hypotheses so that we can test our results against our hypotheses to see if our hypotheses match reality.

So, how do we generate hypotheses in science research?

3.2.1 Hypotheses Always Refer to the Population of People, Plants, or Animals That You Are Studying

The first step is to understand that our hypotheses always refer to the *population* of people, plants, or animals under study.

For example, if we are interested in studying a species of noxious weed found along highways of southern South Dakota, we would select various sections of highways and estimate the number of weeds found in these sections, these sections would be used as our sample. This sample would be used in generalizing our findings for all of the highways in southern South Dakota.

All of the highways in southern South Dakota would be the *population* that we are interested in studying, while the particular sections of highways in our study are called the *sample* from this population.

Since our sample sizes typically contain only a portion of the highways, we are interested in the results of our sample *only insofar as the results of our sample can be "generalized" to the population in which we are really interested.*

That is why our hypotheses always refer to the population, and never to the sample of people, plants, animals, or events in our study.

You will recall from Chap. 1 that we used the symbol: $\overline{X}$ to refer to the mean of the sample we use in our research study (See Sect. 1.1).

We will use the symbol: μ (the Greek letter "mu") to refer to the *population mean*.

In testing our hypotheses, we are trying to decide which one of two competing hypotheses *about the population mean* we should accept given our data set.

3.2.2   The Null Hypothesis and the Research (Alternative) Hypothesis

These two hypotheses are called the *null hypothesis* and the *research hypothesis*.

Statistics textbooks typically refer to the *null hypothesis* with the notation: H_0.

The *research hypothesis* is typically referred to with the notation: H_1, and it is sometimes called the *alternative hypothesis*.

Let's explain first what is meant by the null hypothesis and the research hypothesis:

(1) *The null hypothesis is what we accept as true unless we have compelling evidence that it is not true.*
(2) *The research hypothesis is what we accept as true whenever we reject the null hypothesis as true.*

This is similar to our legal system in America where we assume that a supposed criminal is innocent until he or she is proven guilty in the eyes of a jury. Our null hypothesis is that this defendant is innocent, while the research hypothesis is that he or she is guilty.

In the great state of Missouri, every license plate has the state slogan: "Show me." This means that people in Missouri think of themselves as not gullible enough to accept everything that someone says as true unless that person's actions indicate the truth of his or her claim. In other words, people in Missouri believe strongly that a person's actions speak much louder than that person's words.

Since both the null hypothesis and the research hypothesis cannot both be true, the task of hypothesis testing using statistical formulas is to decide which one you will accept as true, and which one you will reject as true (Schuenemeyer and Drew 2011).

Sometimes in science research a series of rating scales is used to measure people's attitudes toward a company, toward one of its products, or toward their intention-to-buy that company's products. These rating scales are typically five-point, seven-point, or ten-point scales, although other scale values are often used as well.

3.2.2.1   Determining the Null Hypothesis and the Research Hypothesis When Rating Scales Are Used

Here is a typical example of a seven-point scale in science education for parents of eighth grade pupils at the end of a school year (see Fig. 3.6):

Overall, how satisfied are you with the quality of the academic programs offered by your son's or daughter's school?						
1	2	3	4	5	6	7
Very dissatisfied						Very satisfied

Fig. 3.6 Example of a Rating Scale Item for Parents of Eighth Graders (Practical Example)

So, how do we decide what to use as the null hypothesis and the research hypothesis whenever rating scales are used?

Objective: To decide on the null hypothesis and the research hypothesis whenever rating scales are used

In order to make this determination, we will use a simple rule:

Rule: Whenever rating scales are used, we will use the "middle" of the scale as the null hypothesis and the research hypothesis.

In the above example, since 4 is the number in the middle of the scale (i.e., three numbers are below it, and three numbers are above it), our hypotheses become:

Null hypothesis : $\mu = 4$

Research hypothesis : $\mu \neq 4$

In the above rating scale example, if the result of our statistical test for this one attitude scale item indicates that our population mean is "close to 4," we say that we accept the null hypothesis that the parents of eighth grade pupils were neither satisfied nor dissatisfied with the quality of the science program offered by their son's or daughter's school.

In the above example, *if the result of our statistical test indicates that the population mean is significantly different from 4*, we reject the null hypothesis and accept the research hypothesis *by stating either that*:

"Parents of eighth grade pupils were significantly satisfied with the quality of the science program offered by their son's or daughter's school" (this is true whenever our sample mean is significantly greater than our expected population mean of 4).

or

"*Parents of eighth grade pupils were significantly dissatisfied with the quality of the science program offered by their son's or daughter's school*" (this is accepted as true whenever our sample mean is significantly less than our expected population mean of 4).

Both of these conclusions cannot be true. We accept one of the hypotheses as "true" based on the data set in our research study, and the other one as "not true" based on our data set.

The job of the research scientist, then, is to decide which of these two hypotheses, the null hypothesis or the research hypothesis, he or she will accept as true given the data set in the research study.

Let's try some examples of rating scales so that you can practice figuring out what the null hypothesis and the research hypothesis are for each rating scale.

In the spaces in Fig. 3.7, write in the null hypothesis and the research hypothesis for the rating scales:

1. State University is an excellent university.

1	2	3	4	5
Strongly Disagree	Disagree	Undecided	Agree	Strongly Agree

Null hypothesis: μ = _____

Research hypothesis: μ $\neq$ _____

2. How would you rate the quality of teaching in the Environmental Science department at State University?

poor   1   2   3   4   5   6   7   excellent

Null hypothesis: μ = _____

Research hypothesis: μ $\neq$ _____

3. How would you rate the quality of faculty in the Environmental Science department at State University?

1 very poor	2	3	4	5	6	7	8	9	10 very good

Null hypothesis: μ = _____

Research hypothesis: μ $\neq$ _____

Fig. 3.7 Examples of Rating Scales for Determining the Null Hypothesis and the Research Hypothesis

How did you do?

Here are the answers to these three questions:

1. The null hypothesis is $\mu = 3$, and the research hypothesis is $\mu \neq 3$ on this five-point scale (i.e., the "middle" of the scale is 3).
2. The null hypothesis is $\mu = 4$, and the research hypothesis is $\mu \neq 4$ on this seven-point scale (i.e., the "middle" of the scale is 4).
3. The null hypothesis is $\mu = 5.5$, and the research hypothesis is $\mu \neq 5.5$ on this ten-point scale (i.e., the "middle" of the scale is 5.5 since there are 5 numbers below 5.5 and 5 numbers above 5.5).

As another example, suppose Texas Parks and Wildlife uses a four-point scale in its post-hunting satisfaction survey. The results of this survey are used to determine the number of licenses issued for wildlife management the following hunting season. The scale is as follows:

1 = Not So Good
2 = Average
3 = Very Good
4 = Great

On this scale, the null hypothesis is: $\mu = 2.5$ and the research hypothesis is: $\mu \neq 2.5$ because there are two numbers below 2.5, and two numbers above 2.5 on that rating scale.

Now, let's discuss the seven STEPS of hypothesis testing for using the confidence interval about the mean.

3.2.3 The Seven Steps for Hypothesis Testing Using the Confidence Interval About the Mean

> Objective: To learn the seven steps of hypothesis testing using the confidence interval about the mean

There are seven basic steps of hypothesis testing for this statistical test.

3.2.3.1 STEP 1: State the Null Hypothesis and the Research Hypothesis

If you are using numerical scales in your survey, you need to remember that these hypotheses refer to the "middle" of the numerical scale. For example, if you are using seven-point scales with 1 = poor and 7 = excellent, these hypotheses would refer to the middle of these scales and would be:

Null hypothesis H_0 : $\mu = 4$
Research hypothesis H_1 : $\mu \neq 4$

3.2.3.2   STEP 2: Select the Appropriate Statistical Test

In this chapter, we are studying the confidence interval about the mean, and so we will select that test.

3.2.3.3   STEP 3: Calculate the Formula for the Statistical Test

You will recall (see Sect. 3.1.5) that the formula for the confidence interval about the mean is:

$$\overline{X} \pm t \text{ s.e.} \tag{3.2}$$

We discussed the procedure for computing this formula for the confidence interval about the mean using Excel earlier in this chapter, and the steps involved in using that formula are:

1. Use Excel's =COUNT function to find the sample size.
2. Use Excel's =AVERAGE function to find the sample mean, $\overline{X}$.
3. Use Excel's =STDEV function to find the standard deviation, STDEV.
4. Find the standard error of the mean (s.e.) by dividing the standard deviation (STDEV) by the square root of the sample size, n.
5. Use Excel's TINV function to find the lower limit of the confidence interval.
6. Use Excel's TINV function to find the upper limit of the confidence interval.

3.2.3.4   *STEP 4: Draw a Picture of the Confidence Interval About the Mean, Including the Mean, the Lower Limit of the Interval, the Upper Limit of the Interval, and the Reference Value Given in the Null Hypothesis, H_0* **We will explain Step 4 later in the chapter.**

3.2.3.5   STEP 5: Decide on a Decision Rule

(a) *If the reference value is inside the confidence interval, accept the null hypothesis, H_0*
(b) *If the reference value is outside the confidence interval, reject the null hypothesis, H_0, and accept the research hypothesis, H_1*

3.2.3.6 STEP 6: State the Result of Your Statistical Test

There are two possible results when you use the confidence interval about the mean, and only one of them can be accepted as "true." So your result would be one of the following:

Either: Since the reference value is inside the confidence interval, *we accept the null hypothesis, H₀*

   Or: Since the reference value is outside the confidence interval, *we reject the null hypothesis, H₀ , and accept the research hypothesis, H₁*

3.2.3.7 STEP 7: State the Conclusion of Your Statistical Test in Plain English!

In practice, this is more difficult than it sounds because you are trying to summarize the result of your statistical test in simple English that is both concise and accurate so that someone who has never had a statistics course (such as your boss, perhaps) can understand the conclusion of your test. This is a difficult task, and we will give you lots of practice doing this last and most important step throughout this book.

> Objective: To write the conclusion of the confidence interval about the mean test

Let's set some basic rules for stating the conclusion of a hypothesis test.

Rule #1: Whenever you reject H₀ and accept H₁, you must use the word "significantly" in the conclusion to alert the reader that this test found an important result.

Rule #2: Create an outline in words of the "key terms" you want to include in your conclusion so that you do not forget to include some of them.

Rule #3: Write the conclusion in plain English so that the reader can understand it even if that reader has never taken a statistics course.

Let's practice these rules using the Chevy Impala Excel spreadsheet that you created earlier in this chapter, but first we need to state the hypotheses for that car.

If General Motors wants to claim that the Chevy Impala gets 28 miles per gallon on a billboard ad, the hypotheses would be:

$H_0 : \quad \mu = 28\,\text{mpg}$

$H_1 : \quad \mu \neq 28\,\text{mpg}$

You will remember that the reference value of 28 mpg was inside the 95% confidence interval about the mean for your data, so we would accept H_0 for the Chevy Impala that the car does get 28 mpg.

Objective: To state the result when you accept H_0

Result: Since the reference value of 28 mpg is inside the confidence interval, we accept the null hypothesis, H_0

Let's try our three rules now:

Objective: To write the conclusion when you accept H_0

Rule #1: Since the reference value was inside the confidence interval, we cannot use the word "significantly" in the conclusion. This is a basic rule we are using in this chapter for every problem.

Rule #2: The key terms in the conclusion would be:

– Chevy Impala
– reference value of 28 mpg

Rule #3: The Chevy Impala did get 28 mpg.

The process of writing the conclusion when you accept H_0 is relatively straight-forward since you put into words what you said when you wrote the null hypothesis.

However, the process of stating the conclusion when you reject H_0 and accept H_1 is more difficult, so let's practice writing that type of conclusion with three practice case examples:

Objective: To write the result and conclusion when you reject H_0

CASE #1: Suppose that an ad in *The Wall Street Journal* claimed that the Honda Accord Sedan gets 34 miles per gallon. The hypotheses would be:

H_0 : $\mu = 34 \, \text{mpg}$
H_1 : $\mu \neq 34 \, \text{mpg}$

Suppose that your research yields the following confidence interval:

30	31	32	34
lower	Mean	upper	Ref.
limit		limit	Value

Result: *Since the reference value is outside the confidence interval, we reject the null hypothesis and accept the research hypothesis*

The three rules for stating the conclusion would be:

Rule #1: We must include the word "significantly" since the reference value of 34 is outside the confidence interval.

Rule #2: The key terms would be:

  – Honda Accord Sedan
  – significantly
  – either "more than" or "less than"
  – and probably closer to

Rule #3: The Honda Accord Sedan got significantly less than 34 mpg, and it was probably closer to 31 mpg.

Note that this conclusion says that the mpg was less than 34 mpg because the sample mean was only 31 mpg. Note, also, that when you find a significant result by rejecting the null hypothesis, *it is not sufficient to say only: "significantly less than 34 mpg,"* because that does not tell the reader "how much less than 34 mpg" the sample mean was from 34 mpg. To make the conclusion clear, you need to add: "probably closer to 31 mpg" since the sample mean was only 31 mpg.

CASE #2: The National Association of Environmental Professionals (NAEP) is a scholarly environmental society dedicated to the maintenance and protection of the natural and human environment. The NAEP hosts an annual 5-day Conference. Suppose that the NAEP wanted to use an Internet Survey to evaluate the annual conference based on responses from the participants. Let's suppose that you have been asked to perform the data analyses for the returned surveys, and that you want to practice your data analysis skills on the hypothetical Item #15 given in Fig. 3.8:

Item #15:	How likely are you to recommend to colleagues that they attend next year's annual meeting of the NAEP?						
1	2	3	4	5	6	7
very unlikely						very likely

Fig. 3.8 Example of Item #15 of the NAEP Survey

The hypotheses for this one item would be:

$H_0 : \quad \mu = 4$

$H_1 : \quad \mu \neq 4$

Essentially, the null hypothesis equal to 4 states that if the obtained mean score for this question is not significantly different from 4 on the rating scale, then attendees, overall, were neither likely nor unlikely to recommend to colleagues that they attend next year's annual conference.

Suppose that your analysis produced the following confidence interval for this item on the survey.

1.8	2.8	3.8	4
lower	Mean	upper	Ref.
limit		limit	Value

Result: Since the reference value is outside the confidence interval, we reject the null hypothesis and accept the research hypothesis.

Rule #1: You must include the word "significantly" since the reference value is outside the confidence interval.

Rule #2: The key terms would be:

- attendees
- Internet survey
- significantly
- NAEP annual meeting this year
- either likely or unlikely (since the result is significant)
- recommend to colleagues
- attend next year's annual meeting of the NAEP

Rule #3: Attendees at this year's annual meeting of the NAEP were significantly unlikely to recommend to colleagues that they attend next year's annual meeting of the NAEP.

Note that you need to use the word "unlikely" since the sample mean of 2.8 was on the unlikely side of the middle of the rating scale.

CASE #3: The National Association of Environmental Professionals (NAEP) publishes a journal *Environmental Practice* which includes articles dealing with Ecology/Environment. This journal is peer-reviewed and published by the Cambridge University Press. Suppose that the NAEP wanted to evaluate the quality of articles in this journal and has sent out an Internet Survey to its members. Suppose, further, that you have been asked to do the data analysis for the returned surveys, and that you have decided to test your Excel skills on a hypothetical Item #8 of the survey (see Fig. 3.9).

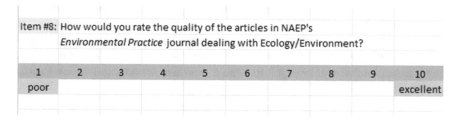

Fig. 3.9 Hypothetical Example of Item #8 of the NAEP Internet Survey

This item would have the following hypotheses:

H_0 : $\mu = 5.5$

H_1 : $\mu \neq 5.5$

Suppose that your research produced the following confidence interval for this item on the survey:

5.5	5.7	5.8	5.9
Ref. Value	lower limit	Mean	upper limit

Result: Since the reference value is outside the confidence interval, we reject the null hypothesis and accept the research hypothesis

The three rules for stating the conclusion would be:

Rule #1: You must include the word "significantly" since the reference value is outside the confidence interval.

Rule #2: The key terms would be:

- Members of the NAEP
- *Environmental Practice*
- significantly
- rated the quality of articles
- Internet survey
- dealing with Ecology/Environment
- either "positive" or "negative" (we will explain this)

Rule #3: Members of the NAEP rated the quality of articles in *Environmental Practice* dealing with Ecology/Environment in an Internet survey as significantly positive.

Note two important things about this conclusion above: (1) people when speaking English do not normally say "significantly excellent" since something is either excellent or is not excellent without any modifier, and (2) since the mean rating of the quality of the articles dealing with Ecology/Environment (5.8) was significantly greater than 5.5 on the positive side of the scale, we would say "significantly positive" to indicate this fact.

If you want a more detailed explanation of the confidence interval about the mean, see Hoshmand (1998).

The three practice problems at the end of this chapter will give you additional practice in stating the conclusion of your result, and this book will include many more examples that will help you to write a clear and accurate conclusion to your research findings.

3.3 Alternative Ways to Summarize the Result of a Hypothesis Test

It is important for you to understand that in this book we are summarizing a hypothesis test in one of two ways: (1) We accept the null hypothesis, or (2) We reject the null hypothesis and accept the research hypothesis. We are consistent in the use of these words so that you can understand the concept underlying hypothesis testing.

However, there are many other ways to summarize the result of a hypothesis test, and all of them are correct theoretically, even though the terminology differs. If you are taking a course with a professor who wants you to summarize the results of a statistical test of hypotheses in language which is different from the language we are using in this book, do not panic! If you understand the concept of hypothesis testing as described in this book, you can then translate your understanding to use the terms that your professor wants you to use to reach the same conclusion to the hypothesis test.

Statisticians and professors of science statistics all have their own language that they like to use to summarize the results of a hypothesis test. There is no one set of words that these statisticians and professors will ever agree on, and so we have chosen the one that we believe to be easier to understand in terms of the concept of hypothesis testing.

To convince you that there are many ways to summarize the results of a hypothesis test, we present the following quotes from prominent statistics and research books to give you an idea of the different ways that are possible.

3.3.1 Different Ways to Accept the Null Hypothesis

The following quotes are typical of the language used in statistics and research books *when the null hypothesis is accepted*:

"The null hypothesis is not rejected." (Black 2010, p. 310)
"The null hypothesis cannot be rejected." (McDaniel and Gates 2010, p. 545)
"The null hypothesis ... claims that there is no difference between groups." (Salkind 2010, p. 193)
"The difference is not statistically significant." (McDaniel and Gates 2010, p. 545)

" . . . the obtained value is not extreme enough for us to say that the difference between Groups 1 and 2 occurred by anything other than chance." (Salkind 2010, p. 225)

"If we do not reject the null hypothesis, we conclude that there is not enough statistical evidence to infer that the alternative (hypothesis) is true." (Keller 2009, p. 358)

"The research hypothesis is not supported." (Zikmund and Babin 2010, p. 552)

3.3.2 Different Ways to Reject the Null Hypothesis

The following quotes are typical of the quotes used in statistics and research books *when the null hypothesis is rejected*:

"The null hypothesis is rejected." (McDaniel and Gates 2010, p. 546)

"If we reject the null hypothesis, we conclude that there is enough statistical evidence to infer that the alternative hypothesis is true." (Keller 2009, p. 358)

"If the test statistic's value is inconsistent with the null hypothesis, we reject the null hypothesis and infer that the alternative hypothesis is true." (Keller 2009, p. 348)

"Because the observed value . . . is greater than the critical value . . ., the decision is to reject the null hypothesis." (Black 2010, p. 359)

"If the obtained value is more extreme than the critical value, the null hypothesis cannot be accepted." (Salkind 2010, p. 243)

"The critical t-value . . . must be surpassed by the observed t-value if the hypothesis test is to be statistically significant . . . ." (Zikmund and Babin 2010, p. 567)

"The calculated test statistic . . . exceeds the upper boundary and falls into this rejection region. The null hypothesis is rejected." (Weiers 2011, p. 330)

You should note that all of the above quotes are used by statisticians and professors when discussing the results of a hypothesis test, and so you should not be surprised if someone asks you to summarize the results of a statistical test using a different language than the one we are using in this book.

3.4 End-of-Chapter Practice Problems

1. The state of Michigan in the USA is known for its great fishing spots in its inland lakes. Suppose that you are working on a research project that identified 230 of these lakes 5 years ago and took a sample of these lakes to determine the average sulfate level (SO_4 in mg/L) in those lakes at that time. Five years ago, the sulfate level averaged 4.65 mg/L in that sample of lakes. The research project wants to determine if the sulfate level in these lakes has changed since then, and you have taken a random sample of these lakes to produce the following hypothetical data in Fig. 3.10.

Fig. 3.10 Worksheet Data
for Chap. 3: Practice
Problem #1

SMALL LAKES IN MICHIGAN
SULFATE LEVELS (SO$_4$ in mg/L)
1.5
1.7
5.5
4.3
3.9
7.4
1.5
2.7
3.3
7.1
6.8
6.3
5.7
5.4
6.2
1.6
1.8
1.9
2.5

(a) To the right of this table, use Excel to find the sample size, mean, standard deviation, and standard error of the mean for the figures. Label your answers. Use number format (two decimal places) for the mean, standard deviation, and standard error of the mean.
(b) Enter the null hypothesis and the research hypothesis onto your spreadsheet.
(c) Use Excel's TINV function to find the 95% confidence interval about the mean for these figures. Label your answers. Use number format (two decimal places).
(d) Enter your *result* onto your spreadsheet.
(e) Enter your *conclusion in plain English* onto your spreadsheet.
(f) Print the final spreadsheet to fit onto one page (if you need help remembering how to do this, see the objectives at the end of Chap. 2 in Sect. 2.4).
(g) On your printout, draw a diagram of this 95% confidence interval by hand.
(h) Save the file as: LAKES3

2. Suppose that you work for an agricultural chemical company and that you have been asked to determine if an inventory level of 12 gallons per business day is sufficient inventory for Brand A in one of its retail outlets during the summer months. You have decided to test your Excel skills before looking at the actual data for gallons sold per business day at this outlet, and the hypothetical data are given in Fig. 3.11.

WEED KILLER INVENTORY FOR BRAND A

Question: "Is an inventory level of 12 gallons per business day sufficient for Brand A?"

DATA (gallons/day)
11
13
12
10
11
13
14
13
12
14
13
13
13
12
14
10

Fig. 3.11 Worksheet Data for Chap. 3: Practice Problem #2

Create an Excel spreadsheet with these data.

(a) Use Excel to the right of the table to find the sample size, mean, standard deviation, and standard error of the mean for these data. Label your answers and use two decimal places for the mean, standard deviation, and standard error of the mean.

(b) Enter the null hypothesis and the research hypothesis for these data on your spreadsheet.

(c) Use Excel's TINV function to find the 95% confidence interval about the mean for these data. Label your answers on your spreadsheet. Use two decimal places for the lower limit and the upper limit of the confidence interval.

(d) Enter the *result* of the test on your spreadsheet.

(e) Enter the *conclusion* of the test in plain English on your spreadsheet.

(f) Print your final spreadsheet so that it fits onto one page (if you need help remembering how to do this, see the objectives at the end of Chap. 2 in Sect. 2.4).

(g) Draw a picture of the confidence interval, including the reference value, onto your spreadsheet.

(h) Save the final spreadsheet as: WEED4

3. Suppose that you have been asked to analyze some environmental impact data from the state of Texas in terms of the amount of SO_2 concentration in the atmosphere in different sites of Texas compared to 3 years ago to see if this concentration (and the air the people who live there breathe) has changed. SO_2 is measured in parts per billion (ppb). Three years ago, when this research was last done, the average concentration of SO_2 in these sites was 120 ppb. Since then, the state has undertaken a comprehensive program to improve the air that people in these sites breathe, and you have been asked to "run the data" to see if any change has occurred.

Is the air that people breathe in these sites now different from the air that people breathed 3 years ago? You have decided to test your Excel skills on a small sample of hypothetical data, and the hypothetical data are given in Fig. 3.12:

CONCENTRATION OF SO_2 IN THE ATMOSPHERE IN PARTS PER BILLION (ppb)

ppb
390
332
186
85
29
135
86
54
28
35
37
28
18
32
24
19
21
35
31
18
20
21
18

Fig. 3.12 Worksheet Data for Chap. 3: Practice Problem #3

Create an Excel spreadsheet with these data.

(a) Use Excel to the right of the table to find the sample size, mean, standard deviation, and standard error of the mean for these data. Label your answers and use two decimal places for the mean, standard deviation, and standard error of the mean.
(b) Enter the null hypothesis and the research hypothesis for these data onto your spreadsheet.
(c) Use Excel's TINV function to find the 95% confidence interval about the mean for these data. Label your answers on your spreadsheet. Use two decimal places for the lower limit and the upper limit of the confidence interval.
(d) Enter the *result* of the test on your spreadsheet.
(e) Enter the *conclusion* of the test in plain English on your spreadsheet.
(f) Print your final spreadsheet so that it fits onto one page (if you need help remembering how to do this, see the objectives at the end of Chap. 2 in Sect. 2.4).
(g) Draw a picture of the confidence interval, including the reference value, onto your spreadsheet.
(h) Save the final spreadsheet as: PARTS3

References

Black, K. Business Statistics: for Contemporary Decision Making (6th ed.). Hoboken, NJ: John Wiley& Sons, Inc., 2010.

Bremer, M. and Doerge, R.W. Statistics at the Bench: A Step-by-Step Handbook for Biologists. Cold Spring Harbor, NY: Cold Spring Harbor Laboratory Press, 2010.

Hoshmand A.R. Statistical methods for environmental and agricultural sciences (2nd ed.). Boca Raton, FL: CRC Press, 1998.

Keller, G. Statistics for Management and Economics (8th ed.). Mason, OH: South-Western Cengage Learning, 2009.

Levine, D.M. Statistics for Managers using Microsoft Excel (6th ed.). Boston, MA: Prentice Hall/ Pearson, 2011.

McDaniel, C. and Gates, R. Marketing Research (8th ed.). Hoboken, NJ: John Wiley & Sons, Inc., 2010.

Salkind, N.J. Statistics for People Who (think they) Hate Statistics (2nd Excel 2007 ed.). Los Angeles, CA: Sage Publications, 2010.

Schuenemeyer J. and Drew L. Statistics for earth and environmental scientists. Hoboken, NJ: John Wiley & Sons, 2011.

Webster R. and Oliver M. Geostatistics for environmental scientists (2nd ed.). Hoboken, NJ: John Wiley & Sons, 2007.

Weiers, R.M. Introduction to Business Statistics (7th ed.). Mason, OH: South-Western Cengage Learning, 2011.

Zikmund, W.G. and Babin, B.J. Exploring Marketing Research (10th ed.). Mason, OH: South-Western Cengage Learning, 2010.

Chapter 4
One-Group t-Test for the Mean

In this chapter, you will learn how to use one of the most popular and most helpful statistical tests in science research: the one-group t-test for the mean.

The formula for the one-group t-test is as follows:

$$t = \frac{\overline{X} - \mu}{S_{\overline{X}}} \text{ where} \tag{4.1}$$

$$\text{s.e.} = S_{\overline{X}} = \frac{S}{\sqrt{n}} \tag{4.2}$$

This formula asks you to take the mean $(\overline{X})$ and subtract the population mean (μ) from it, and then divide the answer by the standard error of the mean (s.e.). The standard error of the mean equals the standard deviation divided by the square root of n (the sample size). If you want to learn more about this test, see Townend (2002) and Hoshmand (1998).

Let's discuss the seven STEPS of hypothesis testing using the one-group t-test so that you can understand how this test is used.

4.1 The Seven STEPS for Hypothesis Testing Using the One-Group t-Test

Objective: To learn the seven steps of hypothesis testing using the one-group t-test

© Springer Nature Switzerland AG 2021
T. J. Quirk et al., *Excel 2019 for Environmental Sciences Statistics*,
Excel for Statistics, https://doi.org/10.1007/978-3-030-66277-6_4

Before you can try out your Excel skills on the one-group t-test, you need to learn the basic steps of hypothesis testing for this statistical test. There are seven steps in this process:

4.1.1   STEP 1: State the Null Hypothesis and the Research Hypothesis

If you are using numerical scales in your survey, you need to remember that these hypotheses refer to the "middle" of the numerical scale. For example, if you are using seven-point scales with $1 = $ poor and $7 = $ excellent, these hypotheses would refer to the middle of these scales and would be:

Null hypothesis H_0 $\mu = 4$
Research hypothesis H_1 $\mu \neq 4$

As a second example, suppose that you worked for Honda Motor Company and that you wanted to place a magazine ad that claimed that the new Honda Fit got 35 miles per gallon (mpg). The hypotheses for testing this claim on actual data would be:

$H_0 : \quad \mu = 35$ mpg
$H_1 : \quad \mu \neq 35$ mpg

4.1.2   STEP 2: Select the Appropriate Statistical Test

In this chapter, we will be studying the one-group t-test, and so we will select that test.

4.1.3   STEP 3: Decide on a Decision Rule for the One-Group t-Test

(a) If the absolute value of t is less than the critical value of t, accept the null hypothesis.
(b) If the absolute value of t is greater than the critical value of t, reject the null hypothesis and accept the research hypothesis.

You are probably saying to yourself: "That sounds fine, but how do I find the absolute value of t?"

4.1.3.1 Finding the Absolute Value of a Number

To do that, we need another objective:

> Objective: To find the absolute value of a number

If you took a basic algebra course in high school, you may remember the concept of "absolute value." In mathematical terms, the absolute value of any number is *always* that number expressed as a positive number.

For example, the absolute value of 2.35 is +2.35.

And the absolute value of minus 2.35 (i.e., −2.35) is also +2.35.

This becomes important when you are using the t-table in Appendix E of this book. We will discuss this table later when we get to Step 5 of the one-group t-test where we explain how to find the critical value of t using Appendix E.

4.1.4 STEP 4: Calculate the Formula for the One-Group t-Test

> Objective: To learn how to use the formula for the one-group t-test

The formula for the one-group t-test is as follows:

$$t = \frac{\overline{X} - \mu}{S_{\overline{X}}} \text{ where} \qquad (4.1)$$

$$\text{s.e.} = S_{\overline{X}} = \frac{S}{\sqrt{n}} \qquad (4.2)$$

This formula makes the following assumptions about the data (Foster et al. 1998): (1) The data are independent of each other (i.e., each person receives only one score), (2) the *population* of the data is normally distributed, and (3) the data have a constant variance (note that the standard deviation is the square root of the variance).

To use this formula, you need to follow these steps:

1. Take the sample mean in your research study and subtract the population mean μ from it (remember that the population mean for a study involving numerical rating scales is the "middle" number in the scale).
2. Then take your answer from the above step and divide your answer by the standard error of the mean for your research study (you will remember that you learned how to find the standard error of the mean in Chap. 1; to find the standard error of the mean, just take the standard deviation of your research study and divide it by the square root of n, where n is the number of people, plants, or animals used in your research study).
3. The number you get after you complete the above step is the value for t that results when you use the formula stated above.

4.1.5   STEP 5: Find the Critical Value of t
in the t-Table in Appendix E

Objective: To find the critical value of t in the t-table in Appendix E

Before we get into an explanation of what is meant by "the critical value of t," let's give you practice in finding the critical value of t by using the t-table in Appendix E.

Keep your finger on Appendix E as we explain how you need to "read" that table.

Since the test in this chapter is called the "one-group t-test," you will use the first column on the left in Appendix E to find the critical value of t for your research study (note that this column is headed: "sample size n").

To find the critical value of t, you go down this first column until you find the sample size in your research study, and then you go to the right and read the critical value of t for that sample size in the critical t column in the table (note that *this is the column that you would use for both the one-group t-test and the 95% confidence interval about the mean*).

For example, if you have 27 people in your research study, the critical value of t is 2.056.

If you have 38 people in your research study, the critical value of t is 2.026.

If you have more than 40 people in your research study, the critical value of t is always 1.96.

Note that the "critical t column" in Appendix E represents the value of t that you need to obtain to be 95% confident of your results as "significant" results.

The critical value of t is the value that tells you whether or not you have found a "significant result" in your statistical test.

The t-table in Appendix E represents a series of "bell-shaped normal curves" (they are called bell-shaped because they look like the outline of the Liberty Bell that you can see in Philadelphia outside of Independence Hall).

The "middle" of these normal curves is treated as if it were zero point on the x-axis (the technical explanation of this fact is beyond the scope of this book, but any good statistics book (e.g., Zikmund and Babin 2010) will explain this concept to you if you are interested in learning more about it).

Thus, values of t that are to the right of this zero point are positive values that use a plus sign before them, and values of t that are to the left of this zero point are negative values that use a minus sign before them. Thus, some values of t are positive, and some are negative.

However, every statistics book that includes a t-table only reprints the *positive* side of the t-curves because the negative side is the mirror image of the positive side; this means that the negative side contains the exact same numbers as the positive side, but the negative numbers all have a minus sign in front of them.

Therefore, to use the t-table in Appendix E, you need to *take the absolute value of the t-value you found when you use the t-test formula* since the t-table in Appendix E only has the values of t that are the positive values for t.

Throughout this book, we are assuming that you want to be 95% confident in the results of your statistical tests. Therefore, the value for t in the t-table in Appendix E tells you whether or not the t-value you obtained when you used the formula for the one-group t-test is within the 95% interval of the t-curve range that t-value would be expected to occur with 95% confidence.

If the t-value you obtained when you used the formula for the one-group t-test is *inside* of the 95% confidence range, we say that the result you found is *not significant* (note that this is equivalent to *accepting the null hypothesis!*).

If the t-value you found when you used the formula for the one-group t-test is *outside* of this 95% confidence range, we say that you have found a *significant result* that would be expected to occur less than 5% of the time (note that this is equivalent to *rejecting the null hypothesis and accepting the research hypothesis*).

4.1.6   STEP 6: State the Result of Your Statistical Test

There are two possible results when you use the one-group t-test, and only one of them can be accepted as "true."

Either: Since the absolute value of t that you found in the t-test formula is *less than the critical value of t* in Appendix E, you accept the null hypothesis.

Or: Since the absolute value of t that you found in the t-test formula is *greater than the critical value of t* in Appendix E, you reject the null hypothesis, and accept the research hypothesis.

4.1.7   STEP 7: State the Conclusion of Your Statistical Test in Plain English!

In practice, this is more difficult than it sounds because you are trying to summarize the result of your statistical test in simple English that is both concise and accurate so that someone who has never had a statistics course (such as your boss, perhaps) can understand the result of your test. This is a difficult task, and we will give you lots of practice doing this last and most important step throughout this book.

If you have read this far, you are ready to sit down at your computer and perform the one-group t-test using Excel on some hypothetical data.

Let's give this a try.

4.2   One-Group t-Test for the Mean

Let's suppose that a local open space park near you had recently created new displays along a nature trail to educate people about the importance of riparian areas for maintaining healthy aquatic ecosystems. Suppose, further, that the organization that manages the park conducted a survey to see how effective their new education messages were with visitors.

The survey contains a number of items, but suppose a hypothetical Item #7 is the one in Fig. 4.1:

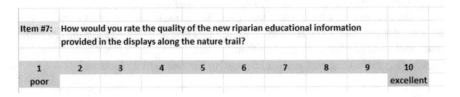

Fig. 4.1 Sample Survey Item for Item #7 of the Riparian Survey (Practical Example)

Suppose further, that you have decided to analyze the data from visitors using the one-group t-test.

Important note: You would need to use this test for each of the survey items separately.

Suppose that the hypothetical data for Item #7 of the Riparian Survey were based on a sample size of 124 visitors who had a mean score on this item of 6.58 and a standard deviation on this item of 2.44.

> Objective: To analyze the data for each question separately using the one-group
>   t-test for each survey item

Create an Excel spreadsheet with the following information:

B11:   Null hypothesis:
B14:   Research hypothesis:

Note: Remember that when you are using a rating scale item, both the null hypothesis and the research hypothesis refer to the "middle of the scale." In the ten-point scale in this example, the middle of the scale is 5.5 since five numbers are below 5.5 (i.e., 1–5) and five numbers are above 5.5 (i.e., 6–10). Therefore, the hypotheses for this rating scale item are:

$H_0 : \mu = 5.5$

$H_1 : \mu \neq 5.5$

B17: n
B20: mean
B23: STDEV

B26: s.e.
B29: critical t
B32: t-test
B36: Result:
B41: Conclusion:

Now, use Excel:

D17: enter the sample size
D20: enter the mean
D23: enter the STDEV (see Fig. 4.2)

Fig. 4.2 Basic Data
Table for Item #7 of the
Riparian Survey

Null hypothesis:		
Research hypothesis:		
n		124
mean		6.58
STDEV		2.44
s.e.		
critical t		
t-test		
Result:		
Conclusion:		

D26   compute the standard error using the formula in Chap. 1
D29:   find the critical t-value of t in the t-table in Appendix E

   Now, enter the following formula in cell D32 to find the t-test result:

=(D20 − 5.5)/D26   (no spaces between)

This formula takes the sample mean (D20) and subtracts the population hypoth-esized mean of 5.5 from the sample mean, and THEN divides the answer by the standard error of the mean (D26). Note that you need to enter D20 − 5.5 with an open-parenthesis *before* D20 and a closed-parenthesis *after* 5.5 so that the *answer of 1.08 is THEN divided by the standard error of 0.22* to get the t-test result of 4.93.

Null hypothesis:	
Research hypothesis:	
n	124
mean	6.58
STDEV	2.44
s.e.	0.22
critical t	1.96
t-test	4.93
Result:	
Conclusion:	

Fig. 4.3   t-test Formula Result for Item #7 of the Riparian Survey

Now, use two decimal places for both the s.e. and the t-test result (see Fig. 4.3).

Now, write the following sentence in D36-D39 to summarize the result of the t-test:

D36:   Since the absolute value of t of 4.93 is
D37:   greater than the critical t of 1.96, we
D38:   reject the null hypothesis and accept
D39:   the research hypothesis.

Lastly, write the following sentence in D41-D43 to summarize the conclusion of the result for Item #7 of the Riparian Survey:

D41:   Visitors rated the quality of the new riparian
D42:   educational information provided in the displays
D43:   along the nature trail as significantly positive.

Save your file as: Riparian4

Important note: We have used the term "significantly positive" because the mean rating of 6.58 is on the positive side of the rating scale. We purposely have not used the term "significantly excellent" because people who speak English do not use that term because something is either excellent or it is not excellent. Therefore, "significantly positive" is a more correct use of the English language in this type of rating scale item.

Important note: You are probably wondering why we entered both the result and the conclusion in separate cells instead of in just one cell. This is because if you enter them in one cell, you will be very disappointed when you print out your final spreadsheet, because one of two things will happen that you will not like: (1) if you print the spreadsheet to fit onto only one page, the result and the conclusion will force the entire spreadsheet to be printed in such small font size that you will be unable to read it, or (2) if you do not print the final spreadsheet to fit onto one page, both the result and the conclusion will "dribble over" onto a second page instead of fitting the entire spread sheet onto one page. In either case, your spreadsheet will not have a "professional look."

Null hypothesis:	μ	=	5.5
Research hypothesis:	μ	≠	5.5
n		124	
mean		6.58	
STDEV		2.44	
s.e.		0.22	
critical t		1.96	
t-test		4.93	
Result:		Since the absolute value of t of 4.93 is greater than the critical t of 1.96, we reject the null hypothesis and accept the research hypothesis.	
Conclusion:		Visitors rated the quality of the new riparian educational information provided in the displays along the nature trail as significantly positive.	

Fig. 4.4 Final Spreadsheet for Item #7 of the Riparian Survey

Print the final spreadsheet so that it fits onto one page as given in Fig. 4.4. Enter the null hypothesis and the research hypothesis by hand on your spreadsheet.
Important note: It is important for you to understand that "technically" the above conclusion in statistical terms should state:

"Visitors rated the quality of the new riparian educational infor-
mation provided in the displays along the nature trail as positive,
and this result was probably not obtained by chance."
However, throughout this book, we are using the term "signifi-
cantly" in writing the conclusion of statistical tests to alert the
reader that the result of the statistical test was probably not a
chance finding, but instead of writing all of those words each
time, we use the word "significantly" as a shorthand to the longer
explanation. This makes it much easier for the reader to understand
the conclusion when it is written "in plain English," instead
of technical, statistical language.

4.3   Can You Use Either the 95% Confidence Interval About the Mean OR the One-Group t-Test When Testing Hypotheses?

You are probably asking yourself:

"It sounds like you could use *either* the 95% confidence interval about the mean *or* the one-group t-test to analyze the results of the types of problems described so far in this book? Is this a correct statement?"

The answer is a resounding: *"Yes!"*

Both the confidence interval about the mean and the one-group t-test are used often in science research on the types of problems described so far in this book. *Both of these tests produce the same result and the same conclusion from the data set!*

Both of these tests are explained in this book because some researchers prefer the confidence interval about the mean test, others prefer the one-group t-test, and still others prefer to use both tests on the same data to make their results and conclusions clearer to the reader of their research reports. Since we do not know which of these tests your researcher prefers, we have explained both of them so that you are competent in the use of both tests in the analysis of statistical data.

Now, let's try your Excel skills on the one-group t-test on these three problems at the end of this chapter.

PHOSPORUS CONCENTRATION (mg/L) IN WASTE WATER EFFLUENT		
CONCENTRATION (mg/L)		
0.0142		
0.0135		
0.0138		
0.0136		
0.0137		
0.0135		
0.0141		
0.0140		
0.0138		
0.0134		
0.0135		
0.0137		
0.0142		
0.0132		
0.0133		

Fig. 4.5 Worksheet Data for Chap. 4: Practice Problem #1

4.4   End-of-Chapter Practice Problems

1. Suppose that the U.S. Environmental Protection Agency (EPA) has set a maximum total phosphorus concentration (mg/L) for wastewater effluent produced by chemical plants to be 0.015 mg/L. Suppose, further, that over a 90-day period, a random sample of wastewater effluent was taken from a specific chemical plant and tested for phosphorus concentration. You have been asked to test your Excel skills on the hypothetical data given in Fig. 4.5.

   (a) Write the null hypothesis and the research hypothesis on your spreadsheet.
   (b) Use Excel to find the sample size, mean, standard deviation, and standard error of the mean to the right of the data set. Use number format (four decimal places) for the mean, standard deviation, and standard error of the mean.
   (c) Enter the critical t from the t-table in Appendix E onto your spreadsheet and label it.
   (d) Use Excel to compute the t-value for these data (use two decimal places) and label it on your spreadsheet.
   (e) Type the result on your spreadsheet, and then type the conclusion in plain

COST OF ON-THE-JOB INJURY CLAIMS	
MONTH	COST ($)
Jan	1,200
Feb	1,875
Mar	2,650
Apr	14,212
May	11,367
Jun	5,126
Jul	6,875
Aug	9,482
Sep	10,265
Oct	19,345
Nov	24,650
Dec	2,311

Fig. 4.6 Worksheet Data for Chap. 4: Practice Problem #2

English on your spreadsheet.
(f) Save the file as: WASTE31

2. Suppose that you work in an environmental science company and that the Human Resource (HR) director has asked you to find out if the company's on-the-job medical claims for this past year were significantly different from these claims from 2 years ago. Two years ago, these claims averaged $16,000 per month. You want to test your Excel skills, and so you have created the hypothetical data in Fig. 4.6.

(a) *On your Excel spreadsheet*, write the null hypothesis and the research hypothesis for these data.
(b) Use Excel to find the *sample size, mean, standard deviation, and standard error of the mean* for these data (use number format, zero decimal places, 1000 separator comma) for the mean, standard deviation, and standard error of the mean).
(c) Use Excel to perform a *one-group t-test* on these data (two decimal places).
(d) On your printout, type the *critical value of t* given in your t-table in Appendix E.
(e) On your spreadsheet, type the *result* of the t-test.
(f) On your spreadsheet, type the *conclusion* of your study in plain English.
(g) Save the file as: INJURY4

DISSOLVED OXYGEN CONTENT(DO) IN MAINE LAKES
DO (mg/L)
4.6
4.4
4.8
6.4
6.5
6.7
6.5
5.6
5.4
5.8
4.9
5.2
5.6
5.7
5.4
4.8

Fig. 4.7 Worksheet Data for Chap. 4: Practice problem #3

3. The state of Maine in the USA is famous for its lakes. There are more than 2,000 named lakes in Maine which are located in the northeastern seaboard of the USA. In addition, there are more than 4,000 other lakes in the state that are greater than one acre in size but have not been named. Dissolved oxygen (DO) is a measure of the quality of the water in a lake. The amount of DO declines as waste is entered into the lakes. Oxygen helps to break down the nutrients in the water, and Burt et al. (2009) state that the DO content of lakes should be 5 milligrams (mg) per liter (L). Suppose that you have collected data on a random sample of named lakes in Maine, and that you want to test your Excel skills on a small sample of these lakes before you try to analyze the data from a much larger sample. The hypothetical data appear in Fig. 4.7.

(a) Write the null hypothesis and the research hypothesis on your spreadsheet.

(b) Use Excel to find the sample size, mean, standard deviation, and standard error of the mean to the right of the data set. Use number format (two decimal places) for the mean, standard deviation, and standard error of the mean.

(c) Enter the critical t from the t-table in Appendix E onto your spreadsheet and label it.

(d) Use Excel to compute the t-value for these data (use two decimal places) and label it on your spreadsheet.

(e) Type the result on your spreadsheet, and then type the conclusion in plain English on your spreadsheet.

(f) Save the file as: MElakes3

References

Burt J, Barber G, Rigby D. Elementary Statistics for Geographers. New York: The Guilford Press; 2009.

Foster, D.P., Stine, R.A., Waterman, R.P. Basic Business Statistics: A Casebook. New York, NY: Springer-Verlag, 1998.

Hoshmand A R. Statistical Methods for Environmental and Agricultural Sciences (2nd ed.). Boca Raton, FL: CRC Press, 1998.

Townend J. Practical Statistics for Environmental and Biological Scientists. Hoboken: John Wiley & Sons, 2002.

Zikmund, W.G. and Babin, B.J. Exploring Marketing Research (10th ed.) Mason, OH: South-Western Cengage Learning, 2010.

Chapter 5
Two-Group t-Test of the Difference of the Means for Independent Groups

Up until now in this book, you have been dealing with the situation in which you have had only one group of people, plants, or animals in your research study and only one measurement "number" on each of these people, plants, or animals. We will now change gears and deal with the situation in which you are measuring two groups instead of only one group.

Whenever you have two completely different groups of people (i.e., no one person is in both groups, but every person is measured on only one variable to produce one "number" for each person), we say that the two groups are "independent of one another." This chapter deals with just that situation and that is why it is called the two-group t-test for independent groups.

The two assumptions underlying the two-group t-test are the following (Zikmund and Babin 2010): (1) both groups are sampled from a normal population, and (2) the variances of the two populations are approximately equal. Note that the standard deviation is merely the square root of the variance. (There are different formulas to use when each person is measured twice to create two groups of data, and this situation is called "dependent," but those formulas are beyond the scope of this book.) This book only deals with two groups that are independent of one another so that no person is in both groups of data.

When you are testing for the difference between the means for two groups, it is important to remember that there are two different formulas that you need to use depending on the sample sizes of the two groups:

(1) Use Formula #1 in this chapter when both of the groups have a sample size greater than 30.
(2) Use Formula #2 in this chapter when either one group, or both groups, have a sample size less than 30.

We will illustrate both of these situations in this chapter.

But, first, we need to understand the steps involved in hypothesis testing when two groups are involved before we dive into the formulas for this test.

© Springer Nature Switzerland AG 2021
T. J. Quirk et al., *Excel 2019 for Environmental Sciences Statistics*,
Excel for Statistics, https://doi.org/10.1007/978-3-030-66277-6_5

5.1   The Nine STEPS for Hypothesis Testing
Using the Two-Group t-Test

Objective: To learn the nine steps of hypothesis testing using two groups of
people, plants, or animals and the two-group t-test

You will see that these steps parallel the steps used in the previous chapter that
dealt with the one-group t-test, but there are some important differences between
the steps that you need to understand clearly before we dive into the formulas for
the two-group t-test.

5.1.1   STEP 1: Name One Group, Group 1,
and the Other Group, Group 2

The formulas used in this chapter will use the numbers 1 and 2 to distinguish
between the two groups. If you define which group is Group 1 and which group is
Group 2, you can use these numbers in your computations without having to write
out the names of the groups.

For example, if you are testing college freshmen males and college freshmen
females, you could call the groups: "College Freshmen Males" and "College
Freshmen Females," but this would require your writing out the words "College
Freshmen Males" and "College Freshmen Females" whenever you wanted to refer to
one of these groups. If you call the College Freshmen Males group, Group 1, and the
College Freshmen Females group, Group 2, this makes it much easier to refer to the
groups because it saves you writing time.

As a second example, you could be comparing flower preference for one type of
hummingbird. Two types of flowers, fuschias and mandevillas, have a vibrant red
color that is a typical attractant found in hummingbird gardens. If you had to write
out the names of the two flowers whenever you wanted to refer to them, it would take
you more time than it would if, instead, you named one flower, Group 1, and the
other flower, Group 2.

Note, also, that it is completely arbitrary which group you call Group 1, and
which Group you call Group 2. You will achieve the same result and the same
conclusion from the formulas however you decide to define these two groups.

5.1.2   STEP 2: Create a Table That Summarizes the Sample Size, Mean Score, and Standard Deviation of Each Group

This step makes it easier for you to make sure that you are using the correct numbers in the formulas for the two-group t-test. If you get the numbers "mixed-up," your entire formula work will be incorrect and you will botch the problem terribly.

For example, the use of fracking in drilling wells is a very contentious topic currently in many parts of the USA. Suppose that you were testing the perceived approval of fracking by college freshman males based on two types of commercials: one with scientists giving the message, and a second commercial with families giving the same message. You could call the groups: "Scientist Message" and "Family Message," but this would require you to write out the words: "Scientist Message" and "Family Message" whenever you wanted to refer to one or both of these groups. If you call the Scientist Message group, Group 1, and the Family Message group, Group 2, this makes it much easier to refer to the groups because it saves you writing time.

Suppose you randomly assigned college freshman males to these two types of commercials such that each group saw only one of the commercials, and then asked these freshmen to rate their perceived approval of fracking on a 100-point scale from $0 = $ poor to $100 = $ excellent. After the research study was completed, suppose that the Scientist Message group had 52 males in it, their mean approval rating was 55 with a standard deviation of 7, while the Family Message group had 57 males in it and their average approval rating was 64 with a standard deviation of 13.

The formulas for analyzing these data to determine if there was a significant difference in the approval rating for freshmen males for these two types of commercials would require you to use six numbers correctly in the formulas: the sample size, the mean, and the standard deviation of each of the two groups. All six of these numbers must be used correctly in the formulas if you are to analyze the data correctly.

If you create a table to summarize these data, a good example of the table, using both Step 1 and Step 2, would be the data presented in Fig. 5.1:

	A	B	C	D	E	F
1						
2						
3		Group	n	Mean	STDEV	
4		1 (name it)				
5		2 (name it)				
6						
7						
8						

Fig. 5.1   Basic Table Format for the Two-group t-test

For example, if you decide to call Group 1 the Scientist Message group and Group 2 the Family Message group, the following table would place the six numbers from your research study into the proper cells of the table as in Fig. 5.2:

◢	A	B	C	D	E	
1						
2						
3		Group	n	Mean	STDEV	
4		1 Scientist Message	52	55	7	
5		2 Family Message	57	64	13	
6						
7						

Fig. 5.2 Results of Entering the Data Needed for the Two-group t-test

You can now use the formulas for the two-group t-test with more confidence that the six numbers will be placed in the proper place in the formulas.

Note that you could just as easily call Group 1 the Family Message group, and Group 2 the Scientist Message group. It makes no difference how you decide to name the two groups; this decision is up to you.

5.1.3   STEP 3: State the Null Hypothesis and the Research Hypothesis for the Two-Group t-Test

If you have completed Step 1 above, this step is very easy because the null hypothesis and the research hypothesis will always be stated in the same way for the two-group t-test. The null hypothesis states that the population means of the two groups (μ) are equal, while the research hypothesis states that the population means of the two groups are not equal. In notation format, this becomes:

$H_0 : \quad \mu_1 = \mu_2$

$H_1 : \quad \mu_1 \neq \mu_2$

You can now see that this notation is much simpler than having to write out the names of the two groups in all of your formulas.

5.1.4   STEP 4: Select the Appropriate Statistical Test

Since this chapter deals with the situation in which you have two groups but only one measurement on each person, plant, or animal in each group, we will use the two-group t-test throughout this chapter.

5.1.5   STEP 5: Decide on a Decision Rule for the Two-Group t-Test

The decision rule is exactly what it was in the previous chapter (see Sect. 4.1.3) when we dealt with the one-group t-test.

(a) If the absolute value of t is less than the critical value of t, accept the null hypothesis.
(b) If the absolute value of t is greater than the critical value of t, reject the null hypothesis and accept the research hypothesis.

Since you learned how to find the absolute value of t in the previous chapter (see Sect. 4.1.3.1), you can use that knowledge in this chapter.

5.1.6   STEP 6: Calculate the Formula for the Two-Group t-Test

Since we are using two different formulas in this chapter for the two-group t-test depending on the sample size in the two groups, we will explain how to use those formulas later in this chapter.

5.1.7   STEP 7: Find the Critical Value of t in the t-Table in Appendix E

In the previous chapter where we were dealing with the one-group t-test, you found the critical value of t in the t-table in Appendix E by finding the sample size for the one group in the first column of the table, and then reading the critical value of t across from it on the right in the "critical t column" in the table (see Sect. 4.1.5). This process was fairly simple once you have had some practice in doing this step.

However, for the two-group t-test, the procedure for finding the critical value of t is more complicated because you have two different groups in your study, and they often have different sample sizes in each group.

To use Appendix E correctly in this chapter, you need to learn how to find the "degrees of freedom" for your study. We will discuss that process now.

5.1.7.1 Finding the Degrees of Freedom (df) for the Two-Group t-Test

> Objective: To find the degrees of freedom for the two-group t-test and to use it to find the critical value of t in the t-table in Appendix E

The mathematical explanation of the concept of the "degrees of freedom" is beyond the scope of this book, but you can find out more about this concept by reading any good statistics book (e.g., Keller 2009). For our purposes, you can easily understand how to find the degrees of freedom and to use it to find the critical value of t in Appendix E. The formula for the degrees of freedom (df) is:

$$\text{degrees of freedom} = \text{df} = n_1 + n_2 - 2 \tag{5.1}$$

In other words, you add the sample size for Group 1 to the sample size for Group 2 and then subtract 2 from this total to get the number of degrees of freedom to use in Appendix E.

Take a look at Appendix E.

Instead of using the first column as we did in the one-group t-test that is based on the sample size, n, of one group, *we need to use the second column of this table (df) to find the critical value of t for the two-group t-test*.

For example, if you had 13 people in Group 1 and 17 people in Group 2, the degrees of freedom would be: $13 + 17 - 2 = 28$, and the critical value of t would be 2.048 *since you look down the second column which contains the degrees of freedom* until you come to the number 28, and then read 2.048 in the "critical t column" in the table to find the critical value of t when df = 28.

As a second example, if you had 52 people in Group 1 and 57 people in Group 2, the degrees of freedom would be: $52 + 57 - 2 = 107$. When you go down the second column in Appendix E for the degrees of freedom, you find that *once you go beyond the degrees of freedom equal to 39, the critical value of t is always 1.96*, and that is the value you would use for the critical t with this example.

5.1.8 STEP 8: State the Result of Your Statistical Test

The result follows the exact same result format that you found for the one-group t-test in the previous chapter (see Sect. 4.1.6):

Either: Since the absolute value of t that you found in the t-test formula is *less than the critical value of t* in Appendix E, you accept the null hypothesis.

Or: Since the absolute value of t that you found in the t-test formula is *greater than the critical value of t* in Appendix E, you reject the null hypothesis and accept the research hypothesis.

5.1.9   STEP 9: State the Conclusion of Your Statistical Test in Plain English

Writing the conclusion for the two-group t-test is more difficult than writing the conclusion for the one-group t-test because you have to decide what the difference was between the two groups.

When you accept the null hypothesis, the conclusion is simple to write: "There is no difference between the two groups in the variable that was measured."

But when you reject the null hypothesis and accept the research hypothesis, you need to be careful about writing the conclusion so that it is both accurate and concise.

Let's give you some practice in writing the conclusion of a two-group t-test.

5.1.9.1   Writing the Conclusion of the Two-Group t-Test When You Accept the Null Hypothesis

> Objective: To write the conclusion of the two-group t-test when you have accepted the null hypothesis

Suppose that college students experienced a one-day environmental education program about the importance of hunting in wildlife management. You have developed a survey and administered it to these students at the end of the day. Item #10 of this survey is given in Fig. 5.3.

Item #10:	As a result of this program, how satisfied are you in your understanding of hunting as an important part of wildlife management?

1	2	3	4	5	6	7	8	9	10
Very Dissatisfied									Very Satisfied

Fig. 5.3   Environmental Education Survey Item #10

Suppose further, that you have decided to analyze the data by comparing Males and Females using the two-group t-test, and that you have decided to call the Males, Group 1, and the Females, Group 2.

Important note: You would need to use this test for each of the survey items separately.

Suppose that the hypothetical data for Item #10 was based on a sample size of 124 males who had a mean score on this item of 6.58 and a standard deviation on this item of 2.44. Suppose that you also had data from 86 females who had a mean score of 6.45 with a standard deviation of 1.86.

We will explain later in this chapter how to produce the results of the two-group t-test using its formulas, but, for now, let's "cut to the chase" and tell you that these data would produce the following in Fig. 5.4:

◢	A	B	C	D	E	F
1						
2						
3		Group	n	Mean	STDEV	
4		1 Males	124	6.58	2.44	
5		2 Females	86	6.45	1.86	
6						
7						

Fig. 5.4 Worksheet Data for Males vs. Females for Item #10 for Accepting the Null Hypothesis

degrees of freedom:	208
critical t:	1.96 (in Appendix E)
t-test formula:	0.44 (when you use your calculator!)
Result:	Since the absolute value of 0.44 is less than the critical t of 1.96, we accept the null hypothesis.
Conclusion:	There was no difference between male and female students in their satisfaction with their understanding of hunting as an important part of wildlife management.

Now, let's see what happens when you reject the null hypothesis (H_0) and accept the research hypothesis (H_1).

5.1.9.2 Writing the Conclusion of the Two-Group t-Test When You Reject the Null Hypothesis and Accept the Research Hypothesis

Objective: To write the conclusion of the two-group t-test when you have rejected the null hypothesis and accepted the research hypothesis

Let's continue with this same example, but with the result that we reject the null hypothesis and accept the research hypothesis.

Let's assume that this time you have data on 85 Males and their mean score on this question was 7.26 with a standard deviation of 2.35. Let's further suppose that you also have data on 48 Females and their mean score on this question was 4.37 with a standard deviation of 3.26. Let Males = Group 1 and Females = Group 2.

Without going into the details of the formulas for the two-group t-test, these data would produce the following result and conclusion based on Fig. 5.5:

	A	B	C	D	E	F
1						
2						
3		Group	n	Mean	STDEV	
4		1 Males	85	7.26	2.35	
5		2 Females	48	4.37	3.26	
6						
7						

Fig. 5.5 Worksheet Data for Item #10 for Obtaining a Significant Difference between Males and Females

Null Hypothesis : $\mu_1 = \mu_2$
Research Hypothesis : $\mu_1 \neq \mu_2$

degrees of freedom: 131
critical t: 1.96 (in Appendix E)
t-test formula: 5.40 (when you use your calculator!)
Result: Since the absolute value of 5.40 is greater than the critical t
                              of 1.96, we reject the null hypothesis and accept the research
                              hypothesis.

Now, you need to compare the ratings of the Males and Females to find out which group had the more positive rating of their environmental education experience using the following rule:

Rule: To summarize the conclusion of the two-group t-test, just compare the means of the two groups, and be sure to use the word "significantly" in your conclusion if you rejected the null hypothesis and accepted the research hypothesis.

A good way to prepare to write the conclusion of the two-group t-test when you are using a rating scale is to place the mean scores of the two groups on a drawing of the scale so that you can visualize the difference of the mean scores. For example, for our environmental education example above, you would draw this "picture" of the scale in Fig. 5.6:

| Item #10: | As a result of this program, how satisfied are you in your understanding of hunting as an important part of wildlife management? |

1	2	3	4	5	6	7	8	9	10
Very Dissatisfied			4.37 Females			7.26 Males			Very Satisfied

Fig. 5.6 Example of Drawing a "Picture" of the Means of the Two Groups on the Rating Scale

This drawing tells you visually that Males had a higher positive rating than Females on this item (7.26 vs. 4.37). *And, since you rejected the null hypothesis and accepted the research hypothesis, you know that you have found a significant difference between the two mean scores.*

So, our conclusion needs to contain the following keywords:

– Males
– Females
– understanding of hunting
– important part of wildlife management
– significantly
– more satisfied *or* less satisfied
– *either* (7.26 vs. 4.37) *or* (4.37 vs. 7.26)

We can use these keywords to write the either of two conclusions which are *logically identical*:

Either Males were significantly more satisfied with their understanding of hunting as an important part of wildlife management than Females (7.26 vs. 4.37).

Or Females were significantly less satisfied with their understanding of hunting as an important part of wildlife management than Males (4.37 vs. 7.26).

Both of these conclusions are accurate, so you can decide which one you want to write. It is your choice.

Also, note that the mean scores in parentheses at the end of these conclusions must match the sequence of the two groups in your conclusion. For example, if you say that: "Males were significantly more satisfied than Females," the end of this conclusion should be: (7.26 vs. 4.37) since you mentioned Males first, and Females second.

Alternately, if you wrote that: "Females were significantly less satisfied than Males," the end of this conclusion should be: (4.37 vs. 7.26) since you mentioned Females first, and Males second.

Putting the two mean scores at the end of your conclusion saves the reader from having to turn back to the table in your research report to find these mean scores to see how far apart the mean scores were.

Now, let's discuss FORMULA #1 that deals with the situation in which both groups have a sample size greater than 30.

Objective: To use FORMULA #1 for the two-group t-test when both groups
have a sample size greater than 30

5.2   Formula #1: Both Groups Have a Sample Size Greater Than 30

The first formula we will discuss will be used when you have two groups with a sample size greater than 30 in each group and one measurement on each member in each group. This formula for the two-group t-test is:

$$t = \frac{\overline{X}_1 - \overline{X}_2}{S_{\overline{X}_1 - \overline{X}_2}} \qquad (5.2)$$

$$\text{where } S_{\overline{X}_1 - \overline{X}_2} = \sqrt{\frac{S_1{}^2}{n_1} + \frac{S_2{}^2}{n_2}} \qquad (5.3)$$

$$\text{and where degrees of freedom} = df = n_1 + n_2 - 2 \qquad (5.1)$$

This formula looks daunting when you first see it, but let's explain some of the parts of this formula:

We have explained the concept of "degrees of freedom" earlier in this chapter, and so you should be able to find the degrees of freedom needed for this formula in order to find the critical value of t in Appendix E.

In the previous chapter, *the formula for the one-group t-test was the following*:

$$t = \frac{\overline{X} - \mu}{S_{\overline{X}}} \qquad (4.1)$$

$$\text{where s.e.} = S_{\overline{X}} = \frac{S}{\sqrt{n}} \qquad (4.2)$$

For the one-group t-test, you found the mean score and subtracted the population mean from it, and then divided the result by the standard error of the mean (s.e.) to get the result of the t-test. You then compared the t-test result to the critical value of t to see if you either accepted the null hypothesis, or rejected the null hypothesis and accepted the research hypothesis.

The two-group t-test requires a different formula because you have two groups, each with a mean score on some variable. You are trying to determine whether to accept the null hypothesis that the *population means of the two groups are equal* (in other words, there is no difference statistically between these two means), or whether the difference between the means of the two groups is "sufficiently large" that you would accept *that there is a significant difference* in the mean scores of the two groups.

The numerator of the two-group t-test asks you to find the difference of the means of the two groups:

$$\overline{X}_1 - \overline{X}_2 \tag{5.4}$$

The next step in the formula for the two-group t-test is to divide the answer you get when you subtract the two means by the standard error of the difference of the two means, and *this is a different standard error of the mean that you found for the one-group t-test because there are two means in the two-group t-test.*

The standard error of the mean when you have two groups is called the "standard error of the difference of the means" between the means of the two groups. This formula looks less scary when you break it down into four steps:

1. Square the standard deviation of Group 1 and divide this result by the sample size for Group 1 (n_1).
2. Square the standard deviation of Group 2 and divide this result by the sample size for Group 2 (n_2).
3. Add the results of the above two steps to get a total score.
4. *Take the square root of this total score* to find the standard error of the difference of the means between the two groups, $S_{\overline{X}_1 - \overline{X}_2} = \sqrt{\frac{S_1^2}{n_1} + \frac{S_2^2}{n_2}}$

This last step is the one that gives students the most difficulty when they are finding this standard error using their calculator because they are in such a hurry to get to the answer that they forget to carry the square root sign down to the last step, and thus get a larger number than they should for the standard error.

5.2.1   An Example of Formula #1 for the Two-Group t-Test

Now, let's use Formula #1 in a situation in which both groups have a sample size greater than 30.

Suppose that a large university offered several sections of Introductory Biology 101 to undergraduates last semester and that it wanted to compare the results of the student evaluation form at the end of the course to see if there were gender differences between Males and Females. Suppose, further, that Item #12 of the student evaluation form is the item given in Fig. 5.7.

Fig. 5.7 Example of a Rating Scale for Item #12 for Introductory Biology 101 (Practical Example)

Suppose you collect these ratings and determine (using your new Excel skills) that the 52 Males in this course had a mean rating of 55 with a standard deviation of 7, while the 57 Females in this course had a mean rating of 64 with a standard deviation of 13.

Note that the two-group t-test does not require that both groups have the same sample size. This is another way of saying that the two-group t-test is "robust" (a fancy term that statisticians like to use).

Your data then produce the following table in Fig. 5.8:

Fig. 5.8 Worksheet Data for Item #12 for Introductory Biology 101

Create an Excel spreadsheet and enter the following information:

B3: Group
B4: 1 Males
B5: 2 Females
C3: n
D3: Mean
E3: STDEV
C4: 52
D4: 55
E4: 7
C5: 57
D5: 64
E5: 13

Now, widen column B so that it is twice as wide as column A, and center the six numbers and their labels in your table (see Fig. 5.9)

◢	A	B	C	D	E	F
1						
2						
3		Group	n	Mean	STDEV	
4		1 Males	52	55	7	
5		2 Females	57	64	13	
6						

Fig. 5.9 Results of Widening Column B and Centering the Numbers in the Cells

B8:   Null hypothesis:
B10:   Research hypothesis:

Since both groups have a sample size greater than 30, you need to use Formula #1 for the t-test for the difference of the means of the two groups.

Let's "break this formula down into pieces" to reduce the chance of making a mistake.

B13:   STDEV1 squared/n1 (note that you square the standard deviation of
       Group 1, and then divide the result by the sample size of Group 1)
B16:   STDEV2 squared/n2
B19:   D13 + D16
B22:   s.e.
B25:   critical t
B28:   t-test
B31:   Result:
B36:   Conclusion: (see Fig. 5.10)

Fig. 5.10 Formula Labels
for the Two-group t-test

Group	n	Mean	STDEV
1 Males	52	55	7
2 Females	57	64	13
Null hypothesis:			
Research hypothesis:			
STDEV1 squared/n1			
STDEV2 squared/n2			
D13 + D16			
s.e.			
critical t			
t-test			
Result:			
Conclusion:			

You now need to compute the values of the above formulas in the following cells:

D13: the result of the formula needed to compute cell B13 (use two decimals)
D16: the result of the formula needed to compute cell B16 (use two decimals)
D19: the result of the formula needed to compute cell B19 (use two decimals)

D22: =SQRT(D19) (use two decimals) (no spaces between)

This formula should give you a standard error (s.e.) of 1.98.

D25: 1.96

(Since df = n1 + n2 − 2, this gives df = 109 − 2 = 107, and the critical t is, therefore, 1.96 in Appendix E.)

D28: =(D4 − D5)/D22 (use two decimals) (no spaces between)

This formula should give you a value for the t-test of: −4.55.

Next, check to see if you have rounded off all figures in D13:D28 to two decimal places (see Fig. 5.11).

Fig. 5.11 Results of the t-test Formula for Item #12 for Introductory Biology 101

Group	n	Mean	STDEV
1 Males	52	55	7
2 Females	57	64	13
Null hypothesis:			
Research hypothesis:			
STDEV1 squared/n1		0.94	
STDEV2 squared/n2		2.96	
D13 + D16		3.91	
s.e.		1.98	
critical t		1.96	
t-test		−4.55	
Result:			
Conclusion:			

Now, write the following sentence in D31 to D34 to summarize the result of the study:

D31: Since the absolute value of −4.55
D32: is greater than the critical t of
D33: 1.96, we reject the null hypothesis
D34: and accept the research hypothesis.

Finally, write the following sentence in D36 to D38 to summarize the conclusion of the study in plain English:

D36: Overall, females rated the quality of Biology 101
D37: this past semester as significantly better than
D38: males (64 vs. 55).

Save your file as: BIOL12

Important note: You are probably wondering why we entered both the result and the conclusion in separate cells instead of in just one cell. This is because if you enter them in one cell, you will be very disappointed when you print out your final spreadsheet, because one of two things will happen that you will not like: (1) if you print the spreadsheet to fit onto only one page, the result and the conclusion will force the entire spreadsheet to be printed in such small font size that you will be unable to read it, or (2) if you do not print the final spreadsheet to fit onto one page, both the result and the conclusion will "dribble over" onto a second page instead of fitting the entire spreadsheet onto one page. In either case, your spreadsheet will not have a "professional look."

Print this file so that it fits onto one page and write by hand the null hypothesis and the research hypothesis on your printout.

The final spreadsheet appears in Fig. 5.12.

Group	n	Mean	STDEV
1 Males	52	55	7
2 Females	57	64	13

Null hypothesis:		μ_1	$=$	μ_2	
Research hypothesis:		μ_1	$\neq$	μ_2	

STDEV1 squared/n1	0.94
STDEV2 squared/n2	2.96
D13 + D16	3.91
s.e.	1.98
critical t	1.96
t-test	-4.55
Result:	Since the absolute value of -4.55 is greater than the critical t of 1.96, we reject the null hypothesis and accept the research hypothesis.
Conclusion:	Overall, females rated the quality of Biology 101 this past semester as significantly better than males (64 vs. 55)

Fig. 5.12 Final Worksheet for Item #12 for Biology 101

If you would like more information about the two-group t-test, see Wheater and Cook (2000).

Now, let's use the second formula for the two-group t-test which we use whenever either one group, or both groups, have a sample size less than 30.

Objective: To use Formula #2 for the two-group t-test when one or both groups have a sample size less than 30

Now, let's look at the case when one or both groups have a sample size less than 30.

5.3 Formula #2: One or Both Groups Have a Sample Size Less Than 30

Suppose that you wanted to see if there was a geographic variation in body length (in mm) of queen honeybees in two different regions of the world (REGION A and REGION B). Suppose, further, that you have been asked to analyze the data from this study and to compare the body lengths of the two regions using the two-group t-test for independent samples. You decide to try out your new Excel skills on a small sample of queen honeybees from each region on the hypothetical data given in Fig. 5.13:

BODY LENGTH (in mm) OF QUEEN HONEYBEES IN TWO REGIONS

REGION A	REGION B
22.56	18.22
20.21	19.21
21.83	19.87
19.02	18.34
20.43	18.65
22.72	19.02
20.65	18.25
18.21	20.73
22.52	19.52
20.23	18.17
	18.75
	18.05

Fig. 5.13 Worksheet Data for Body Length of Queen Honeybees (Practical Example)

Let's call REGION A as Group 1, and REGION B as Group 2.

Null hypothesis : $\mu_1 = \mu_2$
Research hypothesis : $\mu_1 \neq \mu_2$

Note: Since both groups have a sample size less than 30, you need to use Formula #2 in the following steps:

Create an Excel spreadsheet and enter the following information:

A3:   BODY LENGTH (in mm) OF QUEEN HONEYBEES IN TWO REGIONS
B5:   REGION A
C5:   REGION B
B6:   22.56
B15:   20.23
C6:   18.22
C17:   18.05

Now, enter the other figures into this table. Be sure to double-check all of your figures. If you have only one incorrect figure, you will not be able to obtain the correct answer to this problem.

Now, widen columns B and C so that all of the information fits inside the cells. To do this, click on both letters B and C at the top of these columns on your spreadsheet to highlight all of the cells in columns B and C. Then, move the mouse pointer to the right end of the B cell until you get a "cross" sign; then, click on this cross sign and drag the sign to the right until you can read all of the words on your screen. Then, stop clicking! Both Column B and Column C should now be the same width.

Then, center all information in the table except the top title by using the following steps:

Left-click your mouse and highlight cells B5:C17. Then, click on the bottom line, second from the left icon, under "Alignment" at the top-center of Home. All of the information in the table should now be in the center of each cell.

E6:   Null hypothesis:
E8:   Research hypothesis:
E11:   Group
E12:   1 Region A
E13:   2 Region B
F11:   n
G11:   Mean
H11:   STDEV

Your spreadsheet should now look like Fig. 5.14.

BODY LENGTH (in mm) OF QUEEN HONEYBEES IN TWO REGIONS

REGION A	REGION B				
22.56	18.22	Null hypothesis:			
20.21	19.21				
21.83	19.87	Research hypothesis:			
19.02	18.34				
20.43	18.65				
22.72	19.02	Group	n	Mean	STDEV
20.65	18.25	1 Region A			
18.21	20.73	2 Region B			
22.52	19.52				
20.23	18.17				
	18.75				
	18.05				

Fig. 5.14 Queen Honeybees Body Length Worksheet Data for Hypothesis Testing

Now you need to use your Excel skills from Chap. 1 to fill in the sample sizes (n), the Means, and the Standard Deviations (STDEV) in the Table in cells F12:H13. Be sure to double-check your work or you will not be able to obtain the correct answer to this problem if you have only one incorrect figure!

Since both groups have a sample size less than 30, you need to use Formula #2 for the t-test for the difference of the means of two independent samples.

Formula #2 for the two-group t-test is the following:

$$t = \frac{\overline{X}_1 - \overline{X}_2}{S_{\overline{X}_1 - \overline{X}_2}} \tag{5.1}$$

$$\text{where } S_{\overline{X}_1 - \overline{X}_2} = \sqrt{\frac{(n_1 - 1)S_1^2 + (n_2 - 1)S_2^2}{n_1 + n_2 - 2} \left(\frac{1}{n_1} + \frac{1}{n_2} \right)} \tag{5.5}$$

$$\text{and where degrees of freedom} = df = n_1 + n_2 - 2 \tag{5.6}$$

This formula is complicated, and so it will reduce your chance of making a mistake in writing it if you "break it down into pieces" instead of trying to write the formula as one cell entry.

Now, enter these words on your spreadsheet:

E16: $(n1 - 1) \times$ STDEV1 squared
E19: $(n2 - 1) \times$ STDEV2 squared
E22: $n_1 + n_2 - 2$
E25: $1/n_1 + 1/n_2$
E28: s.e.

E31:   critical t
E34:   t-test
B37:   Result:
B40:   Conclusion: (see Fig. 5.15)

BODY LENGTH (in mm) OF QUEEN HONEYBEES IN TWO REGIONS

REGION A	REGION B				
22.56	18.22	Null hypothesis:			
20.21	19.21				
21.83	19.87	Research hypothesis:			
19.02	18.34				
20.43	18.65				
22.72	19.02	Group	n	Mean	STDEV
20.65	18.25	1 Region A	10	20.84	1.55
18.21	20.73	2 Region B	12	18.90	0.82
22.52	19.52				
20.23	18.17				
	18.75	(n1 - 1) x STDEV1 squared			
	18.05				
		(n2 - 1) x STDEV2 squared			
		n1 + n2 - 2			
		1/n1 + 1/n2			
		s.e.			
		critical t			
		t-test			
Result:					
Conclusion:					

Fig. 5.15 Queen Honeybees Body Length Formula Labels for the Two-group t-test

You now need to use your Excel skills to compute the values of the above formulas in the following cells:

H16: the result of the formula needed to compute cell E16 (use two decimals)
H19: the result of the formula needed to compute cell E19 (use two decimals)
H22: the result of the formula needed to compute cell E22
H25: the result of the formula needed to compute cell E25 (use two decimals)
H28: =SQRT(((H16 + H19)/H22)*H25)   (no spaces between)

Note the three open-parentheses after SQRT, and the three closed-parentheses on the right side of this formula. You need three open-parentheses and three closed-parentheses in this formula or the formula will not work correctly.

The above formula gives a standard error of the difference of the means equal to 0.51 (two decimals) in cell H28.

H31: Enter the critical t-value from the t-table in Appendix E in this cell using
      df $= n_1 + n_2 - 2$ to find the critical t-value
H34: =(G12 − G13)/H28 (no spaces between)

Note that you need an open-parenthesis *before G12* and a closed-parenthesis *after G13* so that this answer of 1.94 is *THEN* divided by the standard error of the difference of the means of 0.51, to give a t-test value of 3.77. Use two decimal places for the t-test result (see Fig. 5.16).

BODY LENGTH (in mm) OF QUEEN HONEYBEES IN TWO REGIONS

REGION A	REGION B				
22.56	18.22	Null hypothesis:			
20.21	19.21				
21.83	19.87	Research hypothesis:			
19.02	18.34				
20.43	18.65				
22.72	19.02	Group	n	Mean	STDEV
20.65	18.25	1 Region A	10	20.84	1.55
18.21	20.73	2 Region B	12	18.90	0.82
22.52	19.52				
20.23	18.17				
	18.75	(n1 - 1) x STDEV1 squared			21.50
	18.05				
		(n2 - 1) x STDEV2 squared			7.32
		n1 + n2 - 2			20
		1/n1 + 1/n2			0.18
		s.e.			0.51
		critical t			2.086
		t-test			3.77
Result:					
Conclusion:					

Fig. 5.16 Queen Honeybees Body Length Two-group t-test Formula Results

Now write the following sentence in C37 to C38 to summarize the *result* of the study:

C37:   Since the absolute value of 3.77 is greater than 2.086, we reject the null
C38:   hypothesis and accept the research hypothesis.

Finally, write the following sentence in C40 to C41 to summarize the *conclusion* of the study:

C40:   The body length of queen honeybees in Region A was significantly longer than
C41:   the body length of queen honeybees in Region B (20.84 vs. 18.90).

Save your file as: Honey14

Print the final spreadsheet so that it fits onto one page.
Write the null hypothesis and the research hypothesis by hand on your printout.
The final spreadsheet appears in Fig. 5.17.

BODY LENGTH (in mm) OF QUEEN HONEYBEES IN TWO REGIONS

REGION A	REGION B				
22.56	18.22	Null hypothesis:	μ_1	=	μ_2
20.21	19.21				
21.83	19.87	Research hypothesis:	μ_1	≠	μ_2
19.02	18.34				
20.43	18.65				
22.72	19.02	Group	n	Mean	STDEV
20.65	18.25	1 Region A 10		20.84	1.55
18.21	20.73	2 Region B 12		18.90	0.82
22.52	19.52				
20.23	18.17				
	18.75	(n1 - 1) x STDEV1 squared			21.50
	18.05				
		(n2 - 1) x STDEV2 squared			7.32
		n1 + n2 - 2			20
		1/n1 + 1/n2			0.18
		s.e.			0.51
		critical t			2.086
		t-test			3.77

Result: Since the absolute value of 3.77 is greater than 2.086, we reject the null
            hypothesis and accept the research hypothesis.

Conclusion: The body length of queen honeybees in Region A was significantly longer than
             the body length of queen honeybees in Region B (20.84 vs. 18.90)

Fig. 5.17 Queen Honeybees Body Length Final Spreadsheet

If you would like more information about the two-group t-test, see Hoshmand (1998).

5.4   End-of-Chapter Practice Problems

1. Suppose that you wanted to compare the wing length (in mm) of a species of adult mosquitoes in the northeast region and the southeast region of the USA. You have obtained the cooperation of other biologists in Vermont and New Hampshire in the northeast region, and Kentucky and South Carolina in the southeast region who have shared their data with you from a previous study. In the North, 124 mosquitoes had wings with a mean length 3.2 mm and a standard deviation of 1.2 mm. In the South, 135 mosquitoes had wings with a mean length of 3.4 mm with a standard deviation of 1.3 mm.

   (a) State the null hypothesis and the research hypothesis on an Excel spreadsheet.
   (b) Find the standard error of the difference between the means using Excel.
   (c) Find the critical t-value using Appendix E and enter it on your spreadsheet.
   (d) Perform a t-test on these data using Excel. What is the value of t that you obtain?
      Use three decimal places for all figures in the formula section of your spreadsheet.
   (e) State your result on your spreadsheet.
   (f) State your conclusion in plain English on your spreadsheet.
   (g) Save the file as: Mosquito3

2. Wheater and Cook (2000) discussed an interesting study comparing the amount of sediment in rivers when construction sites are nearby the river (Urban) vs. agricultural land that is nearby the river (Rural). They measured the amount of suspended sediment loads in a section of the river in these two types of sites in milligrams per liter (mg/L). Each river was only used once in the study (i.e., the data are independent samples). The sections of the rivers were similar in size, flow rate, and altitude.

   Suppose that you have been hired as a research assistant in a similar study and that you have been asked to analyze the hypothetical data given in Fig. 5.18:

Fig. 5.18 Worksheet Data
for Chap. 5: Practice
Problem #2

SEDIMENT IN RIVERS IN URBAN AND RURAL SITES

Amount of sediment (mg/L)

urban	rural
45	35
62	40
84	55
95	78
55	38
59	42
64	48
94	52
105	70
87	65
76	44
87	38

(a) On your Excel spreadsheet, write the null hypothesis and the research hypothesis.
(b) Create a table that summarizes these data on your spreadsheet and use Excel to find the sample sizes, the means, and the standard deviations of the two groups in this table. Use two decimal places for the means and standard deviations.
(c) Use Excel to find the standard error of the difference of the means (two decimal places).
(d) Use Excel to perform a two-group t-test. What is the value of *t* that you obtain (use two decimal places)?
(e) On your spreadsheet, type the *critical value of t* using the t-table in Appendix E.
(f) Type your *result* on the test on your spreadsheet.
(g) Type your *conclusion in plain English* on your spreadsheet.
(h) Save the file as: SEDIMENT3

3. A tree can be described as alive or dead, hardwood or softwood, and in many other ways such as its diameter, height, cubic volume, and so forth. In particular, let's assume that you want to compare two species of White Pine (A and B) in

Fig. 5.19 Worksheet Data
for Chap. 5: Practice
Problem #3

WHITE PINE PULPWOOD VOLUME	
TYPE A	TYPE B
6	7
8	14
9	12
7	13
6	8
10	9
12	10
6	8
7	7
9	6
7	12
6	14
7	8
8	14
10	9
8	10

terms of their volume production of pulpwood over a specific time period. You have designed a research study and have staked out thirty-two, square one-acre plots, from which you have selected sixteen plots randomly and planted them with seedlings from White Pine Type A, and the remaining sixteen plots were planted with seedlings from White Pine Type B. After the specified time period, each tree was measured in terms of the pulpwood volume (in cords that were 3 feet wide, by 3 feet high, by 6 feet in length) in each of the thirty-two plots. To test your Excel skills, you have created the hypothetical data given in Fig. 5.19.

(a) State the null hypothesis and the research hypothesis on an Excel spreadsheet.
(b) Find the standard error of the difference between the means using Excel (two decimals).
(c) Find the critical t-value using Appendix E and enter it on your spreadsheet.
(d) Perform a t-test on these data using Excel. What is the value of t that you obtain (two decimals)?
(e) State your result on your spreadsheet.
(f) State your conclusion in plain English on your spreadsheet.
(g) Save the file as: Pine4

References

Hoshmand, A R. Statistical Methods for Environmental and Agricultural Sciences (2nd ed.). Boca Raton, FL: CRC Press, 1998.

Keller, G. Statistics for Management and Economics (8th ed.). Mason, OH: South-Western Cengage Learning, 2009.

Ofungwu J. Statistical Applications for Environmental Analysis and Risk Assessment. Hoboken, NJ: John Wiley & Sons, 2014.

Wheater C, Cook P. Using Statistics to Understand the Environment. New York, NY: Routledge, 2000.

Wisconsin Department of Health Services. http://www.dhs.wisconsin.gov/eh/ChemFS/fs/PCB.htm (October 29, 2014).

Zikmund, W.G. and Babin, B.J. Exploring Marketing Research (10th ed.). Mason, OH: South-Western Cengage Learning, 2010.

Chapter 6
Correlation and Simple Linear Regression

There are many different types of "correlation coefficients," but the one we will use in this book is the Pearson product-moment correlation which we will call: *r*.

6.1   What Is a "Correlation?"

Basically, a correlation is a number between −1 and +1 that summarizes the relationship between two variables, which we will call X and Y.

A correlation can be either positive or negative. *A positive correlation means that as X increases, Y increases. A negative correlation means that as X increases, Y decreases.* In statistics books, this part of the relationship is called the *direction* of the relationship (i.e., it is either positive or negative).

The correlation also tells us the *magnitude* of the relationship between X and Y. As the correlation approaches closer to +1, we say that the relationship is *strong and positive.*

As the correlation approaches closer to −1, we say that the relationship is *strong and negative.*

A zero correlation means that there is no relationship between X and Y. This means that neither X nor Y can be used as a predictor of the other.

A good way to understand what a correlation means is to see a "picture" of the scatterplot of points produced in a chart by the data points. Let's suppose that you want to know if variable X can be used to predict variable Y. We will place *the predictor variable X on the x-axis* (the horizontal axis of a chart) and *the criterion variable Y on the y-axis* (the vertical axis of a chart). Suppose, further, that you have collected data given in the scatterplots below (see Fig. 6.1 through Fig. 6.6).

Figure 6.1 shows the scatterplot for a perfect positive correlation of *r = +1.0*. This means that you can perfectly predict each y-value from each x-value because the data points move "upward-and-to-the-right" along a perfectly fitting straight line (see Fig. 6.1).

© Springer Nature Switzerland AG 2021   113
T. J. Quirk et al., *Excel 2019 for Environmental Sciences Statistics*,
Excel for Statistics, https://doi.org/10.1007/978-3-030-66277-6_6

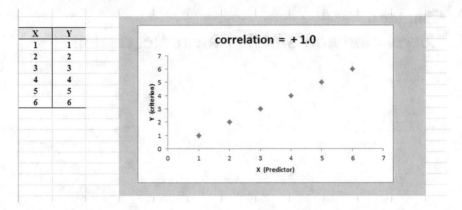

X	Y
1	1
2	2
3	3
4	4
5	5
6	6

Fig. 6.1 Example of a Scatterplot for a Perfect, Positive Correlation (r = +1.0)

Figure 6.2 shows the scatterplot for a moderately positive correlation of *r* = *+.54*. This means that each x-value can predict each y-value moderately well because you can draw a picture of a "football" around the outside of the data points that move upward-and-to-the-right, but not along a straight line (see Fig. 6.2).

X	Y
1	2
2	4
3	3
4	1
5	6
6	5

Fig. 6.2 Example of a Scatterplot for a Moderate, Positive Correlation (r = +.54)

Figure 6.3 shows the scatterplot for a low, positive correlation of *r* = *+.09*. This means that each x-value is a poor predictor of each y-value because the "picture" you could draw around the outside of the data points approaches a circle in shape (see Fig. 6.3).

X	Y
1	2
2	4
3	6
4	1
5	5
6	3

Fig. 6.3 Example of a Scatterplot for a Low, Positive Correlation (r = +.09)

We have not shown a Figure of a zero correlation because it is easy to imagine what it looks like as a scatterplot. A zero correlation of $r = .00$ means that there is no relationship between X and Y and the "picture" drawn around the data points would be a perfect circle in shape, indicating that you cannot use X to predict Y because these two variables are not correlated with one another.

Figure 6.4 shows the scatterplot for a low, negative correlation of $r = -.09$ which means that each X is a poor predictor of Y in an inverse relationship, meaning that as X increases, Y decreases (see Fig. 6.4). In this case, it is a negative correlation because the "football" you could draw around the data points slopes down and to the right.

X	Y
1	3
2	5
3	1
4	6
5	4
6	2

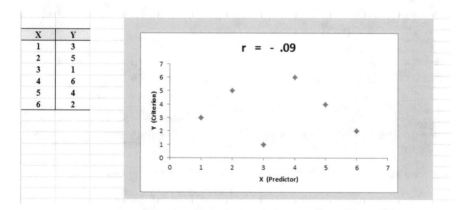

Fig. 6.4 Example of a Scatterplot for a Low, Negative Correlation (r = −.09)

Figure 6.5 shows the scatterplot for a moderate, negative correlation of $r = -.54$ which means that X is a moderately good predictor of Y although there is an inverse relationship between X and Y (i.e., as X increases, Y decreases; see Fig. 6.5). In this case, it is a negative correlation because the "football" you could draw around the data points slopes down and to the right.

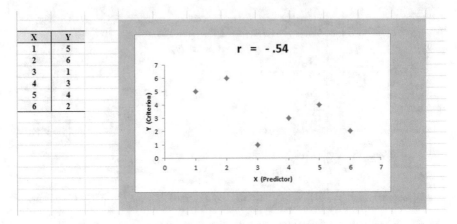

X	Y
1	5
2	6
3	1
4	3
5	4
6	2

Fig. 6.5 Example of a Scatterplot for a Moderate, Negative Correlation (r = −.54)

Figure 6.6 shows a perfect negative correlation of *r* = −*1.0* which means that X is a perfect predictor of Y although in an inverse relationship such that as X increases, Y decreases. The data points fit perfectly along a downward-sloping straight line (see Fig. 6.6).

X	Y
1	6
2	5
3	4
4	3
5	2
6	1

Fig. 6.6 Example of a Scatterplot for a Perfect, Negative Correlation (r = −1.0)

Let's explain the formula for computing the correlation *r* so that you can understand where the number summarizing the correlation came from.

In order to help you to understand *where* the correlation number that ranges from −1.0 to +1.0 comes from, we will walk you through the steps involved to use the formula as if you were using a pocket calculator. This is the one time in this book that we will ask you to use your pocket calculator to find a correlation, but knowing how the correlation is computed step by step will give you the opportunity to understand *how* the formula works in practice.

To do that, let's create a situation in which you need to find the correlation between two variables.

Suppose that you work for a state-of-the-art hydroponic laboratory. You have been asked to conduct a study on the effect of nitrogen on plant growth. You decide to use a predetermined nitrogen solution and to give varying amounts, in milliliters (ml), of the solution to plants in a laboratory greenhouse. In this example, the x-variable (i.e., the predictor variable) is the amount of nitrogen solution (ml). The y-variable (i.e., the criterion variable) is the amount of plant growth in centimeters (cm). To test your Excel skills, you take a random sample of plants from the laboratory greenhouse. The hypothetical data for eight plants appear in Fig. 6.7. *(Note: We are using only one decimal place for these measurements in this example to simplify the mathematical computations.)*

Fig. 6.7 Worksheet Data for Nitrogen Solution and Plant Growth (Practical Example)

	A	B	C
1			
2		X	Y
3	Plant	Nitrogen Solution (ml)	Plant Growth (cm)
4	1	2.8	2.9
5	2	2.5	2.8
6	3	3.1	2.8
7	4	3.5	3.2
8	5	2.4	2.6
9	6	2.6	2.3
10	7	2.4	2.1
11	8	3.6	3.2
12			
13	n	8	8
14	MEAN	2.86	2.74
15	STDEV	0.48	0.39
16			
17			

Notice also that we have used Excel to find the sample size for both variables, X and Y, and the MEAN and STDEV of both variables. (You can practice your Excel skills by seeing if you get these same results when you create an Excel spreadsheet for these data.)

Now, let's use the above table to compute the correlation *r* between nitrogen solution and plant growth using your pocket calculator.

6.1.1   Understanding the Formula for Computing a Correlation

Objective: To understand the formula for computing the correlation r

The formula for computing the correlation r is as follows:

$$r = \frac{\frac{1}{n-1}\Sigma\left(X - \overline{X}\right)\left(Y - \overline{Y}\right)}{S_x S_y} \qquad (6.1)$$

This formula looks daunting at first glance, but let's "break it down into its steps" to understand how to compute the correlation r.

6.1.2   Understanding the Nine Steps for Computing a Correlation, r

Objective: To understand the nine steps of computing a correlation r

The nine steps are as follows:

Step	Computation	Result
1	Find the sample size n by noting the number of plants	8
2	Divide the number 1 by the sample size minus 1 (i.e., 1/7)	0.14286
3	*For each Plant*, take its nitrogen solution and subtract the mean nitrogen solution for the eight plants, and call this $X - \overline{X}$ (For example, for Plant # 6, this would be: 2.6 − 2.86)	−0.26
	Note: With your calculator, this difference is −0.26, but when Excel uses 16 decimal places for every computation, this result could be slightly different for each plant	
4	*For each Plant*, take its growth and subtract the mean Plant growth for the eight plants, and call this $Y - \overline{Y}$ (For example, for Plant #6, this would be: 2.3 − 2.74)	−0.44
5	Then, *for each Plant*, multiply $(X - \overline{X})$ times $(Y - \overline{Y})$ (For example, for Plant # 6 this would be: (−0.26) × (−0.44))	+0.1144
6	Add the results of $(X - \overline{X})$ times $(Y - \overline{Y})$ for the eight plants	+1.09

Steps 1–6 would produce the Excel table given in Fig. 6.8.

◢	A	B	C	D	E	F
1						
2		**X**	**Y**			
3	**Plant**	**Nitrogen Solution (ml)**	**Plant Growth (cm)**	$X - \bar{X}$	$Y - \bar{Y}$	$(X - \bar{X})(Y - \bar{Y})$
4	1	2.8	2.9	-0.06	0.16	-0.01
5	2	2.5	2.8	-0.36	0.06	-0.02
6	3	3.1	2.8	0.24	0.06	0.01
7	4	3.5	3.2	0.64	0.46	0.29
8	5	2.4	2.6	-0.46	-0.14	0.06
9	6	2.6	2.3	-0.26	-0.44	0.11
10	7	2.4	2.1	-0.46	-0.64	0.29
11	8	3.6	3.2	0.74	0.46	0.34
12						- - - - - - -
13	**n**	8	8		**Total**	1.09
14	**MEAN**	2.86	2.74			
15	**STDEV**	0.48	0.39			
16						
17						

Fig. 6.8 Worksheet for Computing the Correlation, r

Notice that when Excel multiplies a minus number by a minus number, the result is a plus number (e.g., for Plant #7: $(-0.46) \times (-0.64) = +0.29$). And when Excel multiplies a minus number by a plus number, the result is a negative number (e.g., for Plant #1: $(-0.06) \times (+0.16) = -0.01$).

Note:Excel computes all computation to 16 decimal places. So, when you check your work with a calculator, you frequently get a slightly different answer than Excel's answer.

For example, when you compute above:

$$(X - \bar{X}) \times (Y - \bar{Y}) \text{ for Plant \#2, your calculator gives :}$$
$$(-0.36) \times (+0.06) = -0.0216 \tag{6.2}$$

As you can see from the table, Excel's answer is -0.02 which is really *more accurate* because Excel uses 16 decimal places for every number, even though only two decimal places are shown in Fig. 6.8.

You should also note that when you do Step 6, you have to be careful to add all of the positive numbers first to get *+1.10* and then add all of the negative numbers second to get *−0.03*, so that when you subtract these two numbers you get *+1.07* as your answer to Step 6. When you do these computations using Excel, this total figure will be *+1.09* because Excel carries every number and computation out to 16 decimal places which is much more accurate than your calculator.

Step		
7	Multiply the answer for step 2 above by the answer for step 6 (0.14286×1.09)	0.1557
8	Multiply the STDEV of X times the STDEV of Y (0.48×0.39)	0.1872
9	Finally, divide the answer from step 7 by the answer from step 8 (0.1557 divided by 0.1872)	+0.83

This number of *0.83* is the correlation between Nitrogen Solution (X) and Plant Growth (Y) for these eight plants. The number +*0.83* means that there is a strong, positive correlation between these two variables. That is, as the amount of nitrogen solution increases, plant growth increases. For a more detailed discussion of correlation, see McCleery et al. (2007).

You could also use the results of the above table in the formula for computing the correlation r in the following way:

$$\text{correlation}\, r = \left[(1/(n-1)) \times \sum (X - \overline{X})(Y - \overline{Y}) \right] / (\text{STDEV}_x \times \text{STDEV}_y)$$

$$\text{correlation}\, r = \left[(1/7) \times 1.09 \right] / \left[(.48) \times (.39) \right]$$

$$\text{correlation} = r = 0.83$$

When you use Excel for these computations, you obtain a slightly different correlation of +0.82 because Excel uses 16 decimal places for all numbers and computations and is, therefore, more accurate than your calculator.

Now, let's discuss how you can use Excel to find the correlation between two variables in a much simpler, and much faster, fashion than using your calculator.

6.2   Using Excel to Compute a Correlation Between Two Variables

Objective: To use Excel to find the correlation between two variables

Suppose that you worked for a car manufacturing company and that you were asked to study the relationship between the weight of 4-door sedans and the fuel consumption they used to drive 150 miles. Suppose, further, that you have obtained 12 sedans, all new models, and have hired drivers to drive 150 miles from Forest Park in St. Louis, Missouri, toward Kansas City, Missouri, on a specified route and at a specified set of speeds. The drivers were all about the same weight.

To test your Excel skills, you have organized the resulting data into a table in which the weight of the cars was measured in thousands of pounds, and the number of gallons of gasoline used in the drive by each car was recorded The hypothetical data appear in Fig. 6.9.

WEIGHT OF 4-DOOR SEDANS VS. NO. OF GALLONS USED TO DRIVE 150 MILES	
Is there a relationship between the weight of a 4-door sedan and the number of gallons used to drive 150 miles?	
WEIGHT (thousands of pounds)	NO. OF GALLONS USED
2.1	5.1
2.3	5.3
2.5	5.2
2.6	5.6
2.7	5.1
3.2	6.1
3.2	6.7
3.4	6.8
3.5	6.8
3.6	6.7
3.8	6.5
4.1	6.9

Fig. 6.9 Worksheet Data for Weight and Number of Gallons Used (Practical Example)

Important note: Note that the weight of the cars is recorded in thousands of pounds, so that a car that weighed 3,500 pounds would be recorded as 3.5 in this table.

You want to determine if there is a *relationship* between the weight of the cars and their fuel consumption, and you decide to use a correlation to determine this relationship. Let's call the weight of the cars the predictor, X, and the number of gallons used, the criterion, Y.

Create an Excel spreadsheet with the following information:

A3: WEIGHT OF 4-DOOR SEDANS VS. NO. OF GALLONS USED TO DRIVE 150 MILES

B5: Is there a relationship between the weight of a 4-door sedan

B6: and the number of gallons used to drive 150 miles?

B8: WEIGHT (thousands of pounds)

C8: NO. OF GALLONS USED

B9: 2.1

C9: 5.1

Next, change the width of Columns B and C so that the information fits inside the cells.

Now, complete the remaining figures in the table given above so that B20 is 4.1 and C20 is 6.9. (Be sure to double-check your figures to make sure that they are correct!) Then, center the information in all of these cells.

A22: n
A23: mean
A24: stdev

Next, define the "name" to the range of data from B9:B20 as: weight

We discussed earlier in this book (see Sect. 1.4.4) how to "name a range of data," but here is a reminder of how to do that:

To give a "name" to a range of data:

Click on the top number in the range of data and drag the mouse down to the bottom number of the range.

For example, to give the name: "weight" to the cells: B9:B20, click on B9, and drag the pointer down to B20 so that the cells B9:B20 are highlighted on your computer screen. Then, click on:

Formulas
Define name (top center of your screen)
weight (enter this in the Name box; see Fig. 6.10)

Fig. 6.10 Dialog Box for Naming a Range of Data as: "weight"

OK

Now, repeat these steps to give the name: gallons to C9:C20

Finally, click on any blank cell on your spreadsheet to "deselect" cells C9:C20 on your computer screen.

Now, complete the data for these sample sizes, means, and standard deviations in columns B and C so that B23 is 3.08, and C24 is 0.75 (use two decimals for the means and standard deviations; see Fig. 6.11).

WEIGHT OF 4-DOOR SEDANS VS. NO. OF GALLONS USED TO DRIVE 150 MILES

Is there a relationship between the weight of a 4-door sedan and the number of gallons used to drive 150 miles?

WEIGHT (thousands of pounds)	NO. OF GALLONS USED
2.1	5.1
2.3	5.3
2.5	5.2
2.6	5.6
2.7	5.1
3.2	6.1
3.2	6.7
3.4	6.8
3.5	6.8
3.6	6.7
3.8	6.5
4.1	6.9
n 12	12
mean 3.08	6.07
stdev 0.63	0.75

Fig. 6.11 Example of Using Excel to Find the Sample Size, Mean, and STDEV

Objective: Find the correlation between weight and gallons used

B26: correlation
C26: =correl(weight,gallons)   (no spaces between) (see Fig. 6.12)

ile	Home	Insert	Page Layout	Formulas	Data	Review	View	Help	♀ Tell me v

| JM | ▼ | ⋮ | × | ✓ | ƒₓ | =correl(weight,gallons) | | | |

A	B	C	D	E

WEIGHT OF 4-DOOR SEDANS VS. NO. OF GALLONS USED TO DRIVE 150 MILES

Is there a relationship between the weight of a 4-door sedan
and the number of gallons used to drive 150 miles?

WEIGHT (thousands of pounds)	NO. OF GALLONS USED
2.1	5.1
2.3	5.3
2.5	5.2
2.6	5.6
2.7	5.1
3.2	6.1
3.2	6.7
3.4	6.8
3.5	6.8
3.6	6.7
3.8	6.5
4.1	6.9

n	12	12
mean	3.08	6.07
stdev	0.63	0.75

	correlation	=correl(weight,gallons)

Fig. 6.12 Example of Using Excel's =correl Function to Compute the Correlation Coefficient

Hit the Enter key to compute the correlation

C26: format this cell to two decimals

Note that the equal sign in =correl(weight,gallons) in C26 tells Excel that you are going to use a formula in this cell.

The correlation between weight (X) and the number of gallons used (Y) is +.91, a very strong positive correlation. This means that you have evidence that there is a strong relationship between these two variables. In effect, the higher the weight, the more gallons needed to drive 150 miles.

Save this file as: GALLONS3

The final spreadsheet appears in Fig. 6.13.

	A	B	C	D	
2					
3	WEIGHT OF 4-DOOR SEDANS VS. NO. OF GALLONS USED TO DRIVE 150 MILES				
4					
5		Is there a relationship between the weight of a 4-door sedan			
6		and the number of gallons used to drive 150 miles?			
7					
8		WEIGHT (thousands of pounds)	NO. OF GALLONS USED		
9		2.1	5.1		
10		2.3	5.3		
11		2.5	5.2		
12		2.6	5.6		
13	GALLONS3	2.7	5.1		
14		3.2	6.1		
15		3.2	6.7		
16		3.4	6.8		
17		3.5	6.8		
18		3.6	6.7		
19		3.8	6.5		
20		4.1	6.9		
21					
22	n	12	12		
23	mean	3.08	6.07		
24	stdev	0.63	0.75		
25					
26		correlation	0.91		

Fig. 6.13 Final Result of Using the =correl Function to Compute the Correlation Coefficient

6.3 Creating a Chart and Drawing the Regression Line onto the Chart

This section deals with the concept of "linear regression." Technically, the use of a simple linear regression model (i.e., the word "simple" means that only one predictor, X, is used to predict the criterion, Y) requires that the data meet the following four assumptions if that statistical model is to be used:

1. The underlying relationship between the two variables under study (X and Y) is *linear* in the sense that a straight line, and not a curved line, can fit among the data points on the chart.
2. The errors of measurement are independent of each other (e.g., the errors from a specific time period are sometimes correlated with the errors in a previous time period).
3. The errors fit a normal distribution of Y-values at each of the X-values.
4. The variance of the errors is the same for all X-values (i.e., the variability of the Y-values is the same for both low and high values of X).

A detailed explanation of these assumptions is beyond the scope of this book, but the interested reader can find a detailed discussion of these assumptions in Levine et al. (2011, pp. 529–530).

Now, let's create a chart summarizing these data.

Important note: Whenever you draw a chart, it is ESSENTIAL that you put the predictor variable (X) on the left, and the criterion variable (Y) on the right in your Excel spreadsheet, so that you know which variable is the predictor variable and which variable is the criterion variable. If you do this, you will save yourself a lot of grief whenever you do a problem involving correlation and simple linear regression using Excel!

You need to understand that in any chart that has one predictor and a criterion that there are really TWO LINES that can be drawn between the data points:

(1) One line uses X as the predictor, and Y as the criterion
(2) A second line uses Y as the predictor, and X as the criterion

This means that you have to be very careful to note in your input data the cells that contain X as the predictor, and Y as the criterion. If you get these cells mixed up and reverse them, you will create the wrong line for your data and you will have botched the problem terribly.

This is why we STRONGLY RECOMMEND IN THIS BOOK that you always put the X data (i.e., the predictor variable) on the LEFT of your table, and the Y data (i.e., the criterion variable) on the RIGHT of your table on your spreadsheet so that you don't get these variables mixed up.

Also note that the correlation, r, will be exactly the same correlation no matter which variable you call the predictor variable and which variable you call the criterion variable. The correlation coefficient just summarizes the relationship between two variables and doesn't care which one is the predictor and which one is the criterion.

Let's suppose that you would like to use weight of the car as the predictor variable, and that you would like to use it to predict the number of gallons needed to drive 150 miles. Since the correlation between these two variables is +.91, this shows that there is a strong, positive relationship and that weight is a good predictor of the number of gallons needed to drive 150 miles.

1. Open the file that you saved earlier in this chapter: GALLONS3

6.3.1   *Using Excel to Create a Chart and the Regression Line Through the Data Points*

> Objective: To create a chart and the regression line summarizing the relationship between weight and gallons used

2. Click and drag the mouse to highlight both columns of numbers (B9:C20), *but do not highlight the labels above the data points.*

Highlight the data set: B9:C20
Insert (top left of screen)
Highlight: Scatter chart icon (immediately above the word: "Charts" at the top center of your screen)
Click on the down arrow on the right of the chart icon
Highlight the top left scatter chart icon (see Fig. 6.14)

Fig. 6.14   Example of Selecting a Scatter Chart

Click on the top left chart to select it
Click on the "+ icon" to the right of the chart (CHART ELEMENTS)
Click on the check mark next to "Chart Title" **and also** next to "Gridlines" to remove these check marks (see Fig. 6.15)

Fig. 6.15 Example of Chart Elements Selected

Click on the box next to "Chart Title" and then click on the arrow to its right. Then, click on: "Above Chart".

Note that the words: "Chart Title" are now in a box at the top of the chart (see Fig. 6.16).

Fig. 6.16 Example of Chart Title Selected

Enter this Chart Title to the right of "Chart f_x" at the top of your screen):
RELATIONSHIP BETWEEN WEIGHT AND NO. OF GALLONS USED (see Fig. 6.17)

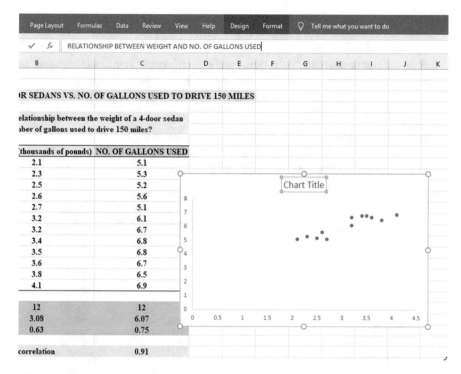

Fig. 6.17   Example of Creating a Chart Title

Hit the enter key to place this title above the chart
Click on *any white space outside of the top title but inside the chart* to "deselect" this
   chart title (see Fig. 6.18).

Fig. 6.18   Example of a Chart Title Inserted Into the Chart

Click on the "+ box" to the right of the chart.

Add a check mark to the left of "Axis Titles" (This will create an "Axis Title" box on
the y-axis of the chart).

Click on the right arrow for: "Axis titles" and then click on: "Primary Horizontal" to
remove the check mark in its box (this will create the y-axis title).

Enter the following y-axis title to the right of f_x at the top of your screen:

NO. OF GALLONS USED

Then, hit the Enter Key to enter this y-axis title to the chart.

Click *inside the chart at the top right corner of the chart to "deselect" the box
around the y-axis title* (see Fig. 6.19).

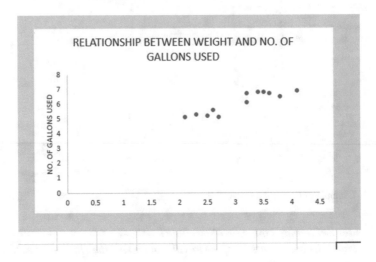

Fig. 6.19 Example of Adding a y-axis Title to the Chart

Click on the "+ box" to the right of the chart.

Highlight: "Axis Titles" and click on its right arrow.

Click on the words: "Primary Horizontal" to add a check mark to its box (this creates
an "Axis Title" box on the x-axis of the chart).

Enter the following x-axis title to the right of f_x at the top of your screen:

WEIGHT (thousands of pounds)

Then, hit the Enter Key to add this x-axis title to the chart.

Click *inside the chart at the top right corner of the chart* to "deselect" the box around
the x-axis title (see Fig. 6.20).

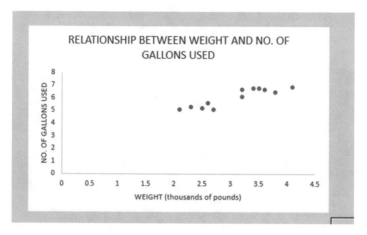

Fig. 6.20   Example of a Chart Title, an x-axis Title, and a y-axis Title

Now, let's draw the regression line onto the chart. This regression line is called the "least-squares regression line" and it is the "best-fitting" straight line through the data points.

6.3.1.1   Drawing the Regression Line Through the Data Points on the Chart

Objective: To draw the regression line through the data points on the chart

Right-click on any one of the data points inside the chart
Highlight: Add Trendline (see Fig. 6.21)

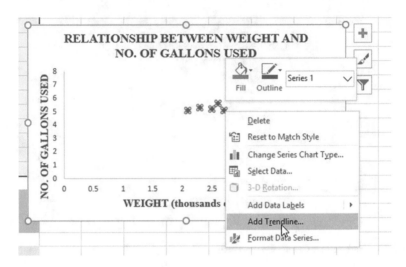

Fig. 6.21 Dialog Box for Adding a Trendline to the Chart

Click on: Add Trendline

Linear (be sure the "Linear" button near the top is selected on the "Format Trendline" dialog box (see Fig. 6.22).

Fig. 6.22 Dialog Box for a Linear Trendline

Click on the X at the top right of the "Format Trendline" dialog box to close this
dialog box.

Now, click on any blank cell *outside the chart* to "deselect" the chart.

Save this file as: GALLONS4

Your spreadsheet should look like the spreadsheet in Fig. 6.23.

Fig. 6.23 Final Chart with the Trendline Fitted Through the Data Points of the Scatterplot

6.3.1.2 Moving the Chart Below the Table in the Spreadsheet

Objective: To move the chart below the table

Left-click your mouse on *any white space to the right of the top title inside the chart*,
keep the left-click down, and drag the chart down and to the left so that the top left
corner of the chart is in cell A29, then take your finger off the left-click of the mouse
(see Fig. 6.24).

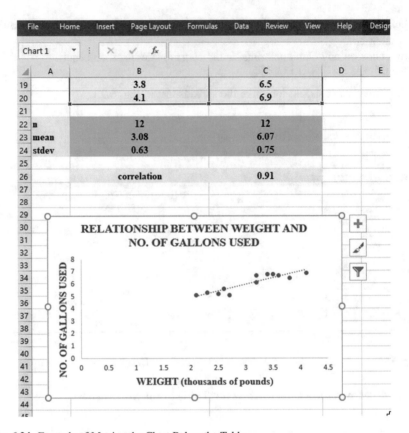

Fig. 6.24 Example of Moving the Chart Below the Table

6.3.1.3   Making the Chart "Longer" So That It Is "Taller"

Objective: To make the chart "longer" so that it is taller

Left-click your mouse on the bottom center of the chart to create an "up-and-down-arrow" sign, hold the left-click of the mouse down and drag the bottom of the chart down to row 48 to make the chart longer, and then take your finger off the mouse.

6.3.1.4   Making the Chart "Wider"

Objective: To make the chart "wider"

Put the pointer at the middle of the right border of the chart to create a "left-to-right arrow" sign, and then left-click your mouse and hold the left-click down while you drag the right border of the chart to the middle of Column H to make the chart wider.

Now, click on any blank cell outside the chart to "deselect" the chart (see Fig. 6.25).

Fig. 6.25 Example of a Chart that is Enlarged to Fit the Cells: A29:H48

Now, click on any blank cell outside the chart to "deselect" the border around the chart.

Save this file as: GALLONS4A

Note: If you printed this spreadsheet now, it is "too big" to fit onto one page, and would "dribble over" onto four pages of printout because the scale needs to be reduced below 100% in order for this worksheet to fit onto only one page. You need to complete these next steps below to print out some, or all, of this spreadsheet.

6.4   Printing a Spreadsheet So That the Table and Chart Fit onto One Page

Objective: To print the spreadsheet so that the table and the chart fit onto one page

Page Layout (top of screen)

Change the scale at the middle icon near the top of the screen "Scale to Fit" by
clicking on the down arrow until it reads "80%" so that the table and the chart will
fit onto one page on your printout (see Fig. 6.26):

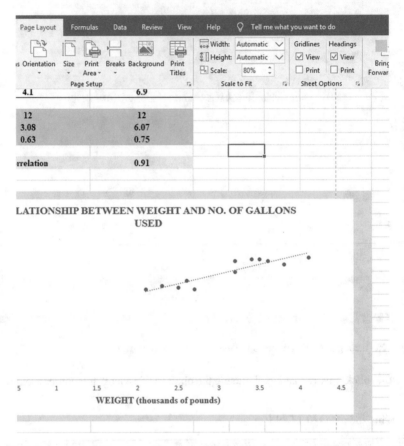

Fig. 6.26 Example of the Page Layout for Reducing the Scale of the Chart to 80% of Normal Size

File
Print
Print (see Fig. 6.27)

WEIGHT OF 4-DOOR SEDANS VS. NO. OF GALLONS USED TO DRIVE 150 MILES

Is there a relationship between the weight of a 4-door sedan and the number of gallons used to drive 150 miles?

WEIGHT (thousands of pounds)	NO. OF GALLONS USED
2.1	5.1
2.3	5.3
2.5	5.2
2.6	5.6
2.7	5.1
3.2	6.1
3.2	6.7
3.4	6.8
3.5	6.8
3.6	6.7
3.8	6.5
4.1	6.9

n	12	12
mean	3.08	6.07
stdev	0.63	0.75
correlation		0.91

RELATIONSHIP BETWEEN WEIGHT AND NO. OF GALLONS USED

Fig. 6.27 Final Spreadsheet of a Table and a Chart (80% Scale to Fit Size)

Save your file as: GALLONS5

6.5 Finding the Regression Equation

The main reason for charting the relationship between X and Y (i.e., weight as X and the number of gallons used as Y in our example) is to see if there is a strong enough relationship between X and Y so that the regression equation that summarizes this relationship can be used to predict Y for a given value of X.

Since we know that the correlation between the weight of the cars and the number of gallons used is +.91, this tells us that it makes sense to use weight to predict the number of gallons used based on past data.

We now need to find that regression equation that is the equation of the "best-fitting straight line" through the data points.

> Objective: To find the regression equation summarizing the relationship between X and Y.

In order to find this equation, we need to check to see if your version of Excel contains the "Data Analysis ToolPak" necessary to run a regression analysis.

6.5.1   Installing the Data Analysis ToolPak into Excel

> Objective: To install the Data Analysis ToolPak into Excel

Since there are currently three versions of Excel in the marketplace (2013, 2016, 2019), we will give a brief explanation of how to install the Data Analysis ToolPak into each of these versions of Excel.

6.5.1.1   Installing the Data Analysis ToolPak into Excel 2019

Open a new Excel spreadsheet

Click on: Data (at the top of your screen)

Look at the top of your monitor screen. Do you see the words: "Data Analysis" at the far right of the screen? If you do, the Data Analysis ToolPak for Excel 2019 was correctly installed when you installed Office 2019, and you should skip ahead to Sect. 6.5.2.

If the words: "Data Analysis" are not at the top right of your monitor screen, then the ToolPak component of Excel 2019 was not installed when you installed Office 2019 onto your computer. If this happens, you need to follow these steps:

File
Options (bottom left of screen)
Note: This creates a dialog box with "Excel Options" at the top left of the box
Add-Ins (on left of screen)
Manage: Excel Add-Ins (at the bottom of the dialog box)
Go (at bottom center of dialog box)
Highlight: Analysis ToolPak (in the Add-Ins dialog box)
Put a check mark to the left of Analysis Toolpak
OK (at the right of this dialog box)
Data

You now should have the words: "Data Analysis" at the top right of your screen to show that this feature has been installed correctly

Note: If these steps do not work, you should try these steps instead:
        File/Options (bottom left)/Add-ins/Analysis ToolPak/Go/
        click to the left of Analysis ToolPak to add a check mark/OK

If you need help doing this, ask your favorite "computer techie" for help.
You are now ready to skip ahead to Sect. 6.5.2.

6.5.1.2   Installing the Data Analysis ToolPak into Excel 2016

Open a new Excel spreadsheet

Click on: Data (at the top of your screen)

Look at the top of your monitor screen. Do you see the words: "Data Analysis" at the far right of the screen? If you do, the Data Analysis ToolPak for Excel 2016 was correctly installed when you installed Office 2016, and you should skip ahead to Sect. 6.5.2.

If the words: "Data Analysis" are not at the top right of your monitor screen, then the ToolPak component of Excel 2016 was not installed when you installed Office 2016 onto your computer. If this happens, you need to follow these steps:

File
Options (bottom left of screen)
Note: This creates a dialog box with "Excel Options" at the top left of the box
Add-Ins (on left of screen)
Manage: Excel Add-Ins (at the bottom of the dialog box)
Go (at bottom center of dialog box)
Highlight: Analysis ToolPak (in the Add-Ins dialog box)
Put a check mark to the left of Analysis Toolpak
OK (at the right of this dialog box)
Data

You now should have the words: "Data Analysis" at the top right of your screen to show that this feature has been installed correctly

Note: If these steps do not work, you should try these steps instead:
        File/Options (bottom left)/Add-ins/Analysis ToolPak/Go/
        click to the left of Analysis ToolPak to add a check mark/OK

If you need help doing this, ask your favorite "computer techie" for help.
You are now ready to skip ahead to Sect. 6.5.2.

6.5.1.3 Installing the Data Analysis ToolPak into Excel 2013

Open a new Excel spreadsheet
   Click on: Data (at the top of your screen)

   Look at the top of your monitor screen. Do you see the words: "Data Analysis" at the far right of the screen? If you do, the Data Analysis ToolPak for Excel 2013 was correctly installed when you installed Office 2013, and you should skip ahead to Sect. 6.5.2.

   If the words: "Data Analysis" are not at the top right of your monitor screen, then the ToolPak component of Excel 2013 was not installed when you installed Office 2013 onto your computer. If this happens, you need to follow these steps:

File
Options (bottom left of screen)
Note: This creates a dialog box with "Excel Options" at the top left of the box
Add-Ins (on left of screen)
Manage: Excel Add-Ins (at the bottom of the dialog box)
Go (at bottom center of dialog box)
Highlight: Analysis ToolPak (in the Add-Ins dialog box)
Put a check mark to the left of Analysis Toolpak
OK (at the right of this dialog box)
Data
You now should have the words: "Data Analysis" at the top right of your screen to
   show that this feature has been installed correctly.

   If you get a prompt asking you for the "installation CD," put this CD in the CD drive and click on: OK

Note: If these steps do not work, you should try these steps instead:
        File/Options (bottom left)/Add-ins/Analysis ToolPak/Go/
        click to the left of Analysis ToolPak to add a check mark/OK

   If you need help doing this, ask your favorite "computer techie" for help.
   You are now ready to skip ahead to Sect. 6.5.2.

6.5.2   Using Excel to Find the SUMMARY OUTPUT of Regression

You have now installed *ToolPak*, and you are ready to find the regression equation for the "best-fitting straight line" through the data points by using the following steps:

Open the Excel file: *GALLONS5* (if it is not already open on your screen)

Note: If this file is already open, and there is a border around the chart, you need to click on any empty cell outside of the chart to deselect the chart.

Now that you have installed *Toolpak*, you are ready to find the regression equation summarizing the relationship between weight and number of gallons used in your data set.

Remember that you gave the name: *weight* to the X data (the predictor), and the name: *gallons* to the Y data (the criterion) in a previous section of this chapter (see Sect. 6.2).

Data (top of screen)
Data analysis (far right at top of screen; see Fig. 6.28)

Fig. 6.28 Example of Using the Data/Data Analysis Function of Excel

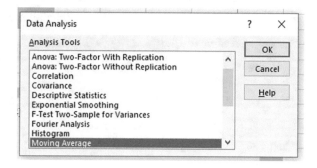

Scroll down the dialog box using the down arrow and click on: Regression (see Fig. 6.29)

Fig. 6.29 Dialog Box for Creating the Regression Function in Excel

OK

Input Y Range: gallons
Input X Range: weight

Click on the "button" to the left of Output Range to select this, and enter A50 in the box as the place on your spreadsheet to insert the Regression analysis in cell A50

OK

The *SUMMARY OUTPUT* should now be in cells: A50:I67
Now, make the columns in the Regression Summary Output section of your spreadsheet *wider* so that you can read all of the column headings clearly.

Now, change the data in the following two cells to Number format (two decimal places):

B53
B66

Next, change this cell to four decimal places: B67

Now, change the format for all other numbers that are in decimal format to number format, three decimal places, and center all numbers within their cells.

Print the file so that it fits onto one page. (*Hint: Change the scale under "Page Layout" to 60% to make it fit.*) Your file should be like the file in Fig. 6.30.

Fig. 6.30 Final Spreadsheet of Correlation and Simple Linear Regression including the SUMMARY OUTPUT for the Data

Save the resulting file as: GALLONS6

Note the following problem with the summary output.

Whoever wrote the computer program for this version of Excel made a mistake and gave the name: "Multiple R" to cell A53.

This is not correct. Instead, cell A53 should say: "correlation r" since this is the notation that we are using for the correlation between X and Y, which is +.91.

You can now use your printout of the regression analysis to find the regression equation that is the best-fitting straight line through the data points.

But first, let's review some basic terms.

6.5.2.1   Finding the y-Intercept, a, of the Regression Line

The point on the y-axis that the regression line would intersect the y-axis if it were extended to reach the y-axis is called the "y-intercept" and *we will use the letter "a" to stand for the y-intercept of the regression line*. The y-intercept on the SUMMARY OUTPUT of Fig. 6.30 is *2.75 and appears in cell B66*. This means that if you were to draw an imaginary line continuing down the regression line toward the y-axis that this imaginary line would cross the y-axis at 2.75. This is why it is called the "y-intercept."

6.5.2.2   Finding the Slope, b, of the Regression Line

The "tilt" of the regression line is called the "slope" of the regression line. It summarizes to what degree the regression line is either above or below a horizontal line through the data points. If the correlation between X and Y were zero, the regression line would be exactly horizontal to the X-axis and would have a zero slope.

If the correlation between X and Y is positive, the regression line would "slope upward to the right" above the X-axis. Since the regression line in Fig. 6.30 slopes upward to the right, the slope of the regression line is +1.0762 as given in cell *B67*. *We will use the notation "b" to stand for the slope of the regression line*. (Note that Excel calls the slope of the line: "X Variable 1" in the Excel printout.)

Since the correlation between weight and gallons used was +*.91*, you can see that the regression line for these data "slopes upward to the right" through the data. Note that the SUMMARY OUTPUT of the regression line in Fig. 6.30 gives a correlation, r, of +*.91* in cell *B53*.

If the correlation between X and Y were negative, the regression line would "slope down to the right" above the X-axis. This would happen whenever the correlation between X and Y is a negative correlation that is between zero and minus one (0 and −1).

6.5.3　Finding the Equation for the Regression Line

To find the regression equation for the straight line that can be used to predict the number of gallons used from the car's weight, we only need two numbers in the SUMMARY OUTPUT in Fig. 6.30: *B66 and B67*.

$$\text{The format for the regression line is}: \quad Y = a + bX \qquad (6.3)$$

where *a = the y-intercept* (2.75 in our example in cell B66)
and *b = the slope of the line* (+1.0762 in our example in cell B67)
　　Therefore, the equation for the best-fitting regression line for our example is:

$$Y = a + b\,X$$

$$\boxed{Y = 2.75 + 1.0762\,X}$$

Remember that Y is the number of gallons used that we are trying to predict, using the weight of the car as the predictor, X.

Let's try an example using this formula to predict the number of gallons used for a car.

6.5.4　Using the Regression Line to Predict the y-Value for a Given x-Value

> Objective: To find the number of gallons predicted for a car that weighed 3,000 pounds (Note: 3,000 pounds, when measured in thousands of pounds, is recorded as 3.0)

Important note: Remember that the weight of the car is in thousands of pounds.

Since the weight is 3,000 pounds (i.e., X = 3.0 in thousands of pounds), substituting this number into our regression equation gives:

$$Y = 2.75 + 1.0762\,(3.0)$$
$$Y = 2.75 + 3.23$$
$$Y = 5.98 \text{ gallons of gas needed to drive 150 miles}$$

Important note: If you look at your chart, if you go directly upward for a weight of 3.0 until you hit the regression line, you see that you hit this line just below 6 on the y-axis to the left when you draw a line horizontal to the x-axis (actually, it is 5.98), the result above for predicting the number of gallons needed for a car weighing 3,000 pounds.

Now, let's do a second example and predict the number of gallons needed if we used a weight of 3,500 pounds. (Remember: 3,500 pounds becomes 3.5 when it is measured in thousands of pounds.)

$$Y = 2.75 + 1.0762\ X$$
$$Y = 2.75 + 1.0762\ (3.5)$$
$$Y = 2.75 + 3.77$$
$$Y = 6.52 \text{ gallons of gas needed to drive 150 miles}$$

Important note: If you look at your chart, if you go directly upward for a weight of 3.5 until you hit the regression line, you see that you hit this line between 6 and 7 on the y-axis to the left (actually it is 6.52), the result above for predicting the number of gallons of gas needed for a car that weighed 3,500 pounds to drive 150 miles.

For a more detailed discussion of regression, see Black (2010) and McKillup and Dyar (2010).

6.6   Adding the Regression Equation to the Chart

Objective: To Add the Regression Equation to the Chart

If you want to include the regression equation within the chart next to the regression line, you can do that, but a word of caution first.

Throughout this book, we are using the regression equation for one predictor and one criterion to be the following:

$$Y = a + b\,X \tag{6.3}$$

where a = y-intercept and b = slope of the line

See, for example, the regression equation in Sect. 6.5.3 where the y-intercept was $a = 2.75$ and the slope of the line was $b = +1.0762$ to generate the following regression equation:

$$Y = 2.75 + 1.0762\ X$$

However, Excel 2019 uses a slightly different regression equation (which is logically identical to the one used in this book) when you add a regression equation to a chart:

$$Y = b\,X + a \tag{6.4}$$

where a = y-intercept and b = slope of the line

Note that this equation is identical to the one we are using in this book with the terms arranged in a different sequence.

For example, we used in Sect. 6.5.3, Excel 2019 would write the regression equation on the chart as:

$$Y = 1.0762\,X + 2.75$$

This is the format that will result when you add the regression equation to the chart using Excel 2019 using the following steps:

Open the file: GALLONS6 *(that you saved in Sect. 6.5.2)*

Click just *inside* the outer border of the chart in the top right corner to add the "border" around the chart in order to "select the chart" for changes you are about to make.

Right-click on any of the data points in the chart.

Highlight: Add Trendline and click on it to select this command.

The "Linear button" near the top of the dialog box will be selected (on its left).

Scroll down this dialog box and click on: Display Equation on chart (near the bottom of the dialog box; see Fig. 6.31).

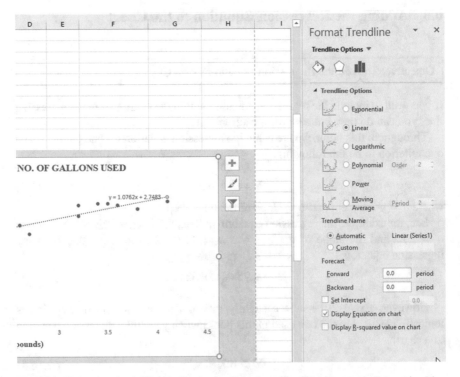

Fig. 6.31 Dialog Box for Adding the Regression Equation to the Chart Next to the Regression Line on the Chart

Click on the X at the top right of the Format Trendline dialog box to remove this box.

Click on any empty cell *outside of the chart* to deselect the chart.

Note that the regression equation on the chart is in the following form next to the regression line on the chart (see Fig. 6.32).

Fig. 6.32 Example of a Chart with the Regression Equation Displayed Next to the Regression Line

$$Y = 1.0762\,X + 2.75$$

Click on any empty cell outside of the chart to deselect the chart.

Now, save this file as: GALLONS7, and print it out so that it fits onto one page.

6.7   How to Recognize Negative Correlations in the SUMMARY OUTPUT Table

Important note: Since Excel does not recognize negative correlations in the SUM-MARY OUTPUT results, but treats all correlations as if they were positive correlations (this was a mistake made by the programmer), you need to be careful to note that there may be a negative correlation between X and Y even if the printout says that the correlation is a positive correlation.

You will know that the correlation between X and Y is a negative correlation when these two things occur:

(1) *THE SLOPE, b, IS A NEGATIVE NUMBER. This can only occur when there is a negative correlation.*
(2) *THE CHART CLEARLY SHOWS A DOWNWARD SLOPE IN THE REGRESSION LINE, which can only occur when the correlation between X and Y is negative.*

6.8   Printing Only Part of a Spreadsheet Instead of the Entire Spreadsheet

> Objective: To print part of a spreadsheet separately instead of printing the entire spreadsheet

There will be many occasions when your spreadsheet is so large in the number of cells used for your data and charts that you only want to print part of the spreadsheet separately so that the print will not be so small that you cannot read it easily.

We will now explain how to print only part of a spreadsheet onto a separate page by using three examples of how to do that using the file, GALLONS7, that you created in Sect. 6.6: (1) printing only the table and the chart on a separate page, (2) printing only the chart on a separate page, and (3) printing only the SUMMARY OUTPUT of the regression analysis on a separate page.

Note: If the file: GALLONS7 is not open on your screen, you need to open it now.

If the "border" is around the outside of the chart, click on any white space outside of the chart to deselect the chart.

Let's describe how to do these three goals with three separate objectives:

6.8.1   *Printing Only the Table and the Chart on a Separate Page*

> Objective: To print only the table and the chart on a separate page

1. Left-click your mouse starting at the top left of the table *in cell A3* and drag the mouse *down and to the right so that all of the table and all of the chart are highlighted in light blue on your computer screen from cell A3 to cell H48* (the light blue cells are called the "selection" cells).
2. File
   Print
   Print Active Sheet (hit the down arrow on the right)
   Print selection
   Print

The resulting printout should contain only the table of the data and the chart resulting from the data.
Then, click on any empty cell in your spreadsheet to deselect the table and chart.

6.8.2   *Printing Only the Chart on a Separate Page*

> Objective: To print only the chart on a separate page

1. Click on any "white space" *just inside the outside border of the chart in the top right corner of the chart* to create the border around all of the borders of the chart in order to "select" the chart.
2. File
   Print
   Print selected chart
   Print selected chart (again)
   Print

The resulting printout should contain only the chart resulting from the data.

Important note: After each time you print a chart by itself on a separate page, you should immediately click on any white space OUTSIDE the chart to remove the border from the chart. When the border is on the borders of the chart, this tells Excel that you want to print only the chart by itself. Do this now!

6.8.3 Printing Only the SUMMARY OUTPUT of the Regression Analysis on a Separate Page

> Objective: To print only the SUMMARY OUTPUT of the regression analysis on a separate page

1. Left-click your mouse at the cell just above SUMMARY OUTPUT in *cell A50* on the left of your spreadsheet and drag the mouse *down and to the right* until all of the regression output is highlighted in dark blue on your screen from A50 to I67.

   Change the "Scale to Fit" to 60% so that the SUMMARY OUTPUT will fit onto one page when you print it out.

2. File

   Print

   Print active sheets (hit the down arrow on the right)

   Print selection

   Print

   The resulting printout should contain only the summary output of the regression analysis on a separate page.

   Finally, click on any empty cell on the spreadsheet to "deselect" the regression table.

6.9 End-of-Chapter Practice Problems

1. What is the relationship between the weight of the car (measured in thousands of pounds) and its city miles per gallon (mpg) in 4-door passenger sedans? Suppose that you wanted to study this question using different models of cars. Analyze the hypothetical data that are given in Fig. 6.33.

Research question:	"What is the relationship between the weight of a 4-door sedan and its miles per gallon (mpg) performance in city driving?"
Weight (1000 lbs)	**City Miles Per Gallon (mpg)**
2.1	32.2
2.4	28.6
3.5	26.7
2.3	28.1
3.4	27.7
4.1	16.2
3.8	20.9
3.6	22.4
4.3	18.4
4.2	15.3

Fig. 6.33 Worksheet Data for Chap. 6: Practice Problem #1

Create an Excel spreadsheet and enter the data.

(a) Create an *XY scatterplot* of these two sets of data such that:

  • top title: RELATIONSHIP BETWEEN WEIGHT AND CITY mpg IN
    4-DOOR SEDANS
  • x-axis title: WEIGHT (1000 lbs)
  • y-axis title: CITY MILES PER GALLON (mpg)
  • move the chart below the table
  • re-size the chart so that it is 7 columns wide and 25 rows long
  • delete the legend
  • delete the gridlines

(b) Create the *least-squares regression line* for these data on the scatterplot.
(c) Use Excel to run the regression statistics to find the *equation for the least-squares regression line* for these data and display the results below the chart on your spreadsheet. Add the regression equation to the chart. Use number format (three decimal places) for the correlation and for the coefficients.
   Print *just the input data and the chart* so that this information fits onto one page in portrait format.
   Then, print *just the regression output table* on a separate page so that it fits onto that separate page in portrait format.
   By hand:
(d) Circle and label the value of the *y-intercept* and the *slope* of the regression line on your printout.
(e) Write the regression equation *by hand* on your printout for these data (use three decimal places for the y-intercept and the slope).
(f) Circle and label the *correlation* between the two sets of scores in the regression analysis summary output table on your printout.
(g) Underneath the regression equation you wrote by hand on your printout, use the regression equation to predict the average city mpg of a 4-door sedan that weighted 2,500 pounds.
(h) *Read from the graph,* the average city mpg you would predict for a 4-door sedan that weighed 3,600 pounds, and write your answer in the space immediately below:

_____

   save the file as: sedan3

2. Permafrost is soil, sediment, or rock that is frozen based on its temperature. The ground must remain at or below zero degrees centigrade for 2 years or more to be called permafrost. It is found at high altitudes, including the Rocky Mountains in the state of Colorado. Permafrost is measured by down-hole depth created by a drill hole in a formation that is used as part of geophysical studies. Suppose that you wanted to study the relationship between down-hole depth (X) and temperature. Suppose that down-hole depth was measured in meters (m) while temperature was measured in degrees centigrade (°C).

Create an Excel spreadsheet and enter the data using DEPTH as the independent (predictor) variable, and TEMPERATURE as the dependent (criterion) variable. You decide to test your Excel skills on a small sample of drill holes using the hypothetical data presented in Fig. 6.34.

DOWN-HOLE DEPTH (meters) VS. TEMPERATURE (degrees centigrade)

DEPTH (m)	TEMPERATURE ($^\circ$C)
0.1	-3.6
0.3	-3.5
0.6	-2.7
0.9	-2.5
1.4	-2.6
2.2	-2.7
3.2	-2.4
4.8	-0.2
6.8	0.0

Fig. 6.34 Worksheet Data for Chap. 6: Practice Problem #2

Create an Excel spreadsheet and enter the data using DEPTH (meters) as the independent variable (predictor) and TEMPERATURE (degrees centigrade) as the dependent variable (criterion).

(a) Create an *XY scatterplot* of these two sets of data such that:

- top title: RELATIONSHIP BETWEEN DOWN-HOLE DEPTH AND TEMPERATURE
- x-axis title: DEPTH (meters)
- y-axis title: TEMPERATURE (degrees centigrade)
- re-size the chart so that it is 7 columns wide and 25 rows long
- delete the legend
- delete the gridlines
- move the chart below the table

(b) Create the *least-squares regression line* for these data on the scatterplot.
(c) Use Excel to run the regression statistics to find the *equation for the least-squares regression line* for these data and display the results below the chart on your spreadsheet. Use number format (two decimal places) for the correlation, r, and for both the y-intercept and the slope of the line. Change all other decimal figures to four decimal places.
(d) Print the input data and the chart so that this information fits onto one page.

(e) Then, print out the regression output table so that this information fits onto a separate page.

By hand:

(1a) Circle and label the value of the *y-intercept* and the *slope* of the regression line onto that separate page.

(2b) *Read from the graph* the temperature you would predict for a *depth of three meters* and write your answer in the space immediately below:

_____

(f) Save the file as: DEPTH3

Answer the following questions using your Excel printout:

1. What is the correlation?
2. What is the y-intercept?
3. What is the slope of the line?
4. What is the regression equation for these data (use two decimal places for the y-intercept and the slope)?
5. Use that regression equation to predict the temperature you would expect for a down-hole depth of two meters.

(Note that this correlation is not the multiple correlation as the Excel table indicates, but is merely the correlation r instead.)

Note that you found a positive correlation of +.94 between depth and temperature. You know that the correlation is a positive correlation for two reasons: (1) the regression line slopes upward and to the right on the chart, signaling a positive correlation, and (2) the slope is +0.53 which also tells you that the correlation is a positive correlation.

But how does Excel treat *negative correlations*?

Important note: Since Excel does not recognize negative correlations in the SUMMARY OUTPUT but treats all correlations as if they were positive correlations, you need to be careful to note when there is a negative correlation between the two variables under study.

You know that the correlation is negative when:

(1) The slope, b, is a negative number which can only occur when there is a negative correlation.

(2) The chart clearly shows a downward slope in the regression line, which can only happen when the correlation is negative.

3. Suppose that you wanted to study the relationship between respiratory problems and air pollution. You have decided to measure air pollution in terms of the level of Sulphur Dioxide (SO_2) in the air (measured in $\mu g/m^3$) during the month of July in a certain city and the average peak flow for women ages 30–40 (measured in ml/s). Peak flow is the maximum rate than a woman can expel air out of her lungs. To test your Excel skills, you have developed the hypothetical data for a city given in Fig. 6.35.

RELATIONSHIP BETWEEN AIR POLLUTION LEVEL AND PEAK FLOW IN ADULT WOMEN (ages 30-40)	
AIR POLLUTION (μ g/m^3)	PEAK FLOW (ml/sec)
21	52
63	43
45	51
103	46
25	48
34	44
39	42
45	41
43	45
48	47
55	51
89	41
96	35
91	32
102	40

Fig. 6.35 Worksheet Data for Chap. 6: Practice Problem #3

Create an Excel spreadsheet and enter the data.

(a) Create an *XY scatterplot* of these two sets of data such that:

- top title: RELATIONSHIP BETWEEN POLLUTION LEVEL AND PEAK FLOW
- x-axis title: POLLUTION
- y-axis title: PEAK FLOW (ml/sec)
- move the chart below the table
- re-size the chart so that it is 7 columns wide and 25 rows long
- delete the legend
- delete the gridlines

(b) Create the *least-squares regression line* for these data on the scatterplot.

(c) Use Excel to run the regression statistics to find the *equation for the least-squares regression line* for these data and display the results below the chart on your spreadsheet. Use number format (two decimal places) for the correlation and for the coefficients.

Print *just the input data and the chart* so that this information fits onto one page in portrait format.

Then, print *just the regression output table* on a separate page so that it fits onto that separate page in portrait format.

By hand:

(d) Circle and label the value of the *y-intercept* and the *slope* of the regression line on your printout.

(e) Write the regression equation *by hand* on your printout for these data (use two decimal places for the y-intercept and the slope).

(f) Circle and label the *correlation* between the two sets of scores in the regression analysis summary output table on your printout.

(g) Underneath the regression equation you wrote by hand on your printout, use the regression equation to predict the average peak flow you would predict for a pollution level of 80.

(h) *Read from the graph,* the average peak flow you would predict for a pollution level of 40 and write your answer in the space immediately below:

_____

Save the file as: Pollution4

References

Black K. Business statistics: for contemporary decision making. 6th ed. Hoboken: John Wiley & Sons, Inc.; 2010.

Levine D, Stephan D, Krehbiel T, Berenson M. Statistics for managers using microsoft excel. 6th ed. Boston: Prentice Hall Pearson; 2011.

McCleery R, Watt T, Hart T. Introduction to statistics for biology. 3rd ed. Boca Raton: Chapman & Hall/CRC; 2007.

McKillup S, Dyar M. Geostatistics explained: an introductory guide for earth scientists. Cambridge: Cambridge University Press; 2010.

Chapter 7
Multiple Correlation and Multiple Regression

There are many times in science when you want to predict a criterion, Y, but you want to find out if you can develop a better prediction model by using *several predictors* in combination (e.g., X_1, X_2, X_3) instead of a single predictor, X.

The resulting statistical procedure is called "multiple correlation" because it uses two or more predictors in combination to predict Y, instead of a single predictor, X. Each predictor is "weighted" differently based on its separate correlation with Y and its correlation with the other predictors. The job of multiple correlation is to produce a regression equation that will weight each predictor differently and in such a way that the combination of predictors does a better job of predicting Y than any single predictor by itself. We will call the multiple correlation: R_{xy}.

You will recall (see Sect. 6.5.3) that the regression equation that predicts Y when only one predictor, X, is used is:

$$Y = a + b\,X \tag{7.1}$$

7.1 Multiple Regression Equation

The multiple regression equation follows a similar format and is:

$$Y = a + b_1 X_1 + b_2 X_2 + b_3 X_3 + etc. \text{ depending on the number of predictors used} \tag{7.2}$$

The "weight" given to each predictor in the equation is represented by the letter "b" with a subscript to correspond to the same subscript on the predictors.

You will remember from Chap. 6 that the correlation, r, ranges from -1 to $+1$. However, the multiple correlation, R_{xy}, only ranges from zero to $+1$. R_{xy} is never a negative number!

© Springer Nature Switzerland AG 2021
T. J. Quirk et al., *Excel 2019 for Environmental Sciences Statistics*,
Excel for Statistics, https://doi.org/10.1007/978-3-030-66277-6_7

Important note: In order to do multiple regression, you need to have installed the "Data Analysis ToolPak" that was described in Chap. 6 (see Sect. 6.5.1). If you did not install this, you need to do so now.

Let's try a practice problem.

Suppose that you have been asked to analyze some data on fruit production from a mid-size fruit farm in Colorado. The farm operates a laboratory greenhouse where it tests new fruit growth under controlled conditions. The farm is able to control the amount of water (milliliters, ml), the amount of light received (minutes, min), and the amount of fertilizer (microliters, μl) a fruit plant receives. Suppose that the farm wants you to determine the relationship between water, light, and fertilizer in their ability to predict the mass of fruit (grams, g) produced. The farm manager has also asked you to look at the amount of fruit produced in relation to growing conditions.

You have decided to use the three measured growing conditions as the predictors $(X_1, X_2,$ and $X_3)$ and the mass of fruit produced as the criterion, Y. To test your Excel skills, you have randomly selected 11 plants and recorded their growing conditions.

Let's use the following notation:

Y	FRUIT PRODUCED
X_1	WATER
X_2	LIGHT RECEIVED
X_3	FERTILIZER

Suppose, further, that you have collected the following hypothetical data summarizing these scores (see Fig. 7.1):

	A	B	C	D	E
1					
2	FRUIT PRODUCED IN RELATION TO GROWING CONDITIONS				
3					
4	Is there a relationship between fruit produced and growing conditions?				
5					
6	FRUIT PRODUCED (g)	WATER (ml)	LIGHT RECEIVED (min)	FERTILIZER (µl)	
7	2.55	250	230	220	
8	3.05	610	240	440	
9	3.55	620	540	530	
10	2.05	420	420	260	
11	2.45	320	520	320	
12	2.95	630	620	620	
13	3.15	650	540	530	
14	3.45	520	580	560	
15	3.30	420	490	630	
16	2.75	330	220	610	
17	3.65	440	570	660	
18					

Fig. 7.1 Worksheet Data for Growing Conditions versus Fruit Produced (Practical Example)

Create an Excel spreadsheet for these data using the following cell reference:

A2:   FRUIT PRODUCED IN RELATION TO GROWING CONDITIONS
A4:   Is there a relationship between fruit produced and growing conditions?
A6:   FRUIT PRODUCED (g)
A7:   2.55
B6:   WATER (ml)
C6:   LIGHT RECEIVED (min)
D6:   FERTILIZER (µl)
D17:   660

Next, change the column width to match the above table and change all FRUIT PRODUCED figures to number format (two decimal places).

Now, fill in the additional data in the chart such that:

A17:   3.65
B17:   440
C17   570

Then, center all numbers in your table

Important note: Be sure to double-check all of your numbers in your table to be sure that they are correct, or your spreadsheets will be incorrect.

Save this file as: FRUIT3

Before we do the multiple regression analysis, we need to try to make one important point very clear:

Important: *When we used one predictor, X, to predict one criterion, Y, we said that you need to make sure that the X variable is ON THE LEFT in your table, and the Y variable is ON THE RIGHT in your table so that you don't get these variables mixed up (see Sect. 6.3).*

However, in multiple regression, you need to follow this rule which is exactly the opposite:

When you use several predictors in multiple regression, it is essential that the criterion you are trying to predict, Y, be ON THE FAR LEFT, and all of the predictors are TO THE RIGHT of the criterion, Y, in your table so that you know which variable is the criterion, Y, and which variables are the predictors. If you make this a habit, you will save yourself a lot of grief.

Notice in the table above, that the criterion Y (FRUIT PRODUCED) is on the far left of the table, and the three predictors (WATER, LIGHT RECEIVED, and FERTILIZER) are to the right of the criterion variable. If you follow this rule, you will be less likely to make a mistake in this type of analysis.

7.2   Finding the Multiple Correlation and the Multiple Regression Equation

Objective: To find the multiple correlation and multiple regression equation using Excel.

You do this by the following commands:

Data
Click on: Data Analysis (far right top of screen)
Regression (scroll down to this in the box; see Fig. 7.2)

Fig. 7.2 Dialog Box for Regression Function

OK

Input Y Range: A6:A17
Input X Range: B6:D17

Note that both the input Y Range and the Input X Range above both include the label at the top of the columns.

Click on the Labels box to *add a check mark* to it (because you have included the column labels in row 6)

Output Range (click on the button to its left, and enter): A20 (see Fig. 7.3)

Fig. 7.3 Dialog Box for Growing Conditions vs. Fruit Produced Data

Important note: Excel automatically assigns a dollar sign $ in front of each column letter and each row number so that you can keep these ranges of data constant for the regression analysis.

OK (see Fig. 7.4 to see the resulting SUMMARY OUTPUT)

FRUIT PRODUCED IN RELATION TO GROWING CONDITIONS

Is there a relationship between fruit produced and growing conditions?

FRUIT PRODUCED (g)	WATER (ml)	LIGHT RECEIVED (min)	FERTILIZER (μl)
2.55	250	230	220
3.05	610	240	440
3.55	620	540	530
2.05	420	420	260
2.45	320	520	320
2.95	630	620	620
3.15	650	540	530
3.45	520	580	560
3.30	420	490	630
2.75	330	220	610
3.65	440	570	660

SUMMARY OUTPUT

Regression Statistics	
Multiple R	0.797651156
R Square	0.636247366
Adjusted R Square	0.48035338
Standard Error	0.361446932
Observations	11

ANOVA

	df	SS	MS	F	Significance F
Regression	3	1.599583719	0.533194573	4.081282	0.057174747
Residual	7	0.91450719	0.130643884		
Total	10	2.514090909			

	Coefficients	Standard Error	t Stat	P-value	Lower 95%	Upper 95%	Lower 95.0%	Upper 95.0%
Intercept	1.53627108	0.468442063	3.279532734	0.013496	0.428581617	2.643960543	0.428581617	2.643960543
WATER (ml)	0.000642945	0.000963026	0.667629207	0.525762	-0.001634251	0.00292014	-0.001634251	0.00292014
LIGHT RECEIVED (min)	0.000264354	0.000889915	0.297055329	0.775046	-0.00183996	0.002368667	-0.00183996	0.002368667
FERTILIZER (μl)	0.00210733	0.000848684	2.4830572	0.042022	0.000100512	0.004114149	0.000100512	0.004114149

Fig. 7.4 Regression SUMMARY OUTPUT of Growing Conditions vs. Fruit Produced Data

Next, format cell B23 in number format (two decimal places).

Next, format the following four cells in Number format (four decimal places):

B36

B37

B38

B39

Change all other decimal figures to two decimal places and center all figures within their cells.

Save the file as: FRUIT4

Now, print the file so that it fits onto one page by changing the scale to 55% size. The resulting regression analysis is given in Fig. 7.5.

FRUIT PRODUCED IN RELATION TO GROWING CONDITIONS

Is there a relationship between fruit produced and growing conditions?

FRUIT PRODUCED (g)	WATER (ml)	LIGHT RECEIVED (min)	FERTILIZER (µl)
2.55	250	230	220
3.05	610	240	440
3.55	620	540	530
2.05	420	420	260
2.45	320	520	320
2.95	630	620	620
3.15	650	540	530
3.45	520	580	560
3.30	420	490	630
2.75	330	220	610
3.65	440	570	660

SUMMARY OUTPUT

Regression Statistics	
Multiple R	0.80
R Square	0.64
Adjusted R Square	0.48
Standard Error	0.36
Observations	11

ANOVA

	df	SS	MS	F	Significance F
Regression	3	1.60	0.53	4.08	0.06
Residual	7	0.91	0.13		
Total	10	2.51			

	Coefficients	Standard Error	t Stat	P-value	Lower 95%	Upper 95%	Lower 95.0%	Upper 95.0%
Intercept	1.5363	0.47	3.28	0.01	0.43	2.64	0.43	2.64
WATER (ml)	0.0006	0.00	0.67	0.53	0.00	0.00	0.00	0.00
LIGHT RECEIVED (min)	0.0003	0.00	0.30	0.78	0.00	0.00	0.00	0.00
FERTILIZER (µl)	0.0021	0.00	2.48	0.04	0.00	0.00	0.00	0.00

Fig. 7.5 Final Spreadsheet for Growing Conditions vs. Fruit Produced Regression Analysis

Once you have the SUMMARY OUTPUT, you can determine the multiple correlation and the regression equation that is the best-fit line through the data points using WATER, LIGHT RECEIVED, and FERTILIZER as the three predictors, and FRUIT PRODUCED as the criterion.

Note on the SUMMARY OUTPUT where it says: "Multiple R." This term is correct since this is the term Excel uses for the multiple correlation, which is +0.80. This means, that from these data, that the combination of WATER, LIGHT RECEIVED, and FERTILIZER together form a very strong positive relationship in predicting FRUIT PRODUCED.

To find the regression equation, *notice the coefficients at the bottom of the SUMMARY OUTPUT*:

Intercept: a (this is the y-intercept)	1.5363
WATER: b_1	0.0006
LIGHT RECEIVED: b_2	0.0003
FERTILIZER: b_3	0.0021

Since the general form of the multiple regression equation is:

$$Y = a + b_1 X_1 + b_2 X_2 + b_3 X_3 \qquad (7.2)$$

we can now write the multiple regression equation for these data:

$$Y = 1.5363 + 0.0006 X_1 + 0.0003\ X_2 + 0.0021 X_3$$

7.3 Using the Regression Equation to Predict FRUIT PRODUCED

> Objective: To find the predicted FRUIT PRODUCED using a WATER Score of 600, a LIGHT RECEIVED Score of 500, and a FERTILIZER Score of 550

Plugging these three numbers into our regression equation gives us:

$$Y = 1.5363 + 0.0006(600) + 0.0003(500) + 0.0021(550)$$
$$Y = 1.5363 + 0.36 + 0.15 + 1.155$$
$$Y = 3.20\ \text{grams of fruit produced}$$

If you want to learn more about the theory behind multiple regression, see Keller (2009) and Hoshmand (1998).

7.4 Using Excel to Create a Correlation Matrix in Multiple Regression

The final step in multiple regression is to find the correlation between all of the variables that appear in the regression equation.

In our example, this means that we need to find the correlation between each of the six pairs of variables:

To do this, we need to use Excel to create a "correlation matrix." This matrix summarizes the correlations between all of the variables in the problem.

> Objective: To use Excel to create a correlation matrix between the four variables in this example

To use Excel to do this, use these steps:
Data (top of screen under "Home" at the top left of screen)

Data Analysis
Correlation (scroll *up* to highlight this formula; see Fig. 7.6)

Fig. 7.6 Dialog Box for
Growing
Conditions vs. Fruit
Produced Correlations

OK

Input range: A6:D17

(Note that this input range includes the labels at the top of the FOUR variables
(FRUIT PRODUCED, WATER, LIGHT RECEIVED, and FERTILIZER) as well
as all of the figures in the original data set.)
Grouped by: Columns
Put a check in the box for:
   Labels in the First Row (since you included the labels at the top of the columns in
   your input range of data above)
Output range (click on the button to its left, and enter): A42 (see Fig. 7.7)

Fig. 7.7   Dialog Box for Input/Output Range for Correlation Matrix

OK
The resulting correlation matrix appears in A42:E46 (See Fig. 7.8).

	A	B	C	D	E	
36	Intercept	1.5363	0.47	3.28	0.01	
37	WATER (ml)	0.0006	0.00	0.67	0.53	
38	LIGHT RECEIVED (min)	0.0003	0.00	0.30	0.78	
39	FERTILIZER (µl)	0.0021	0.00	2.48	0.04	
40						
41						
42		FRUIT PRODUCED (g)	WATER (ml)	LIGHT RECEIVED (min)	FERTILIZER (µl)	
43	FRUIT PRODUCED (g)	1				
44	WATER (ml)	0.510369686	1			
45	LIGHT RECEIVED (min)	0.446857676	0.468105152	1		
46	FERTILIZER (µl)	0.772523347	0.444074496	0.429202393	1	
47						

Fig. 7.8 Resulting Correlation Matrix for Growing Conditions vs. Fruit Produced Data

Next, format all of the numbers in the correlation matrix that are in decimals to two decimals places. And, also, make the columns wider so that all of the labels fit inside their cells. Then, center all the correlations in their cells.

Save this Excel file as: FRUIT5

The final spreadsheet for these scores appears in Fig. 7.9.

FRUIT PRODUCED IN RELATION TO GROWING CONDITIONS

Is there a relationship between fruit produced and growing conditions?

FRUIT PRODUCED (g)	WATER (ml)	LIGHT RECEIVED (min)	FERTILIZER (µl)
2.55	250	230	220
3.05	610	240	440
3.55	620	540	530
2.05	420	420	260
2.45	320	520	320
2.95	630	620	620
3.15	650	540	530
3.45	520	580	560
3.30	420	490	630
2.75	330	220	610
3.65	440	570	660

SUMMARY OUTPUT

Regression Statistics	
Multiple R	0.80
R Square	0.64
Adjusted R Square	0.48
Standard Error	0.36
Observations	11

ANOVA

	df	SS	MS	F	Significance F
Regression	3	1.60	0.53	4.08	0.06
Residual	7	0.91	0.13		
Total	10	2.51			

	Coefficients	Standard Error	t Stat	P-value	Lower 95%	Upper 95%	Lower 95.0%	Upper 95.0%
Intercept	1.5363	0.47	3.28	0.01	0.43	2.64	0.43	2.64
WATER (ml)	0.0006	0.00	0.67	0.53	0.00	0.00	0.00	0.00
LIGHT RECEIVED (min)	0.0003	0.00	0.30	0.78	0.00	0.00	0.00	0.00
FERTILIZER (µl)	0.0021	0.00	2.48	0.04	0.00	0.00	0.00	0.00
FRUIT4								

	FRUIT PRODUCED (g)	WATER (ml)	LIGHT RECEIVED (min)	FERTILIZER (µl)
FRUIT PRODUCED (g)	1			
WATER (ml)	0.51	1		
LIGHT RECEIVED (min)	0.45	0.47	1	
FERTILIZER (µl)	0.77	0.44	0.43	1

Fig. 7.9 Final Spreadsheet for Growing Conditions vs. Fruit Produced Regression and the Correlation Matrix

Note that the number "1" along the diagonal of the correlation matrix means that the correlation of each variable with itself is a perfect, positive correlation of 1.0. *Correlation coefficients are always expressed in just two decimal places.* You are now ready to read the correlation between the six pairs of variables:

The correlation between WATER and FRUIT PRODUCED is:	*+.51*
The correlation between LIGHT RECEIVED and FRUIT PRODUCED is:	*+.45*
The correlation between FERTILIZER and FRUIT PRODUCED is:	*+.77*
The correlation between LIGHT RECEIVED and WATER is:	*+.47*
The correlation between FERTILIZER and WATER is:	*+.44*
The correlation between FERTILIZER and LIGHT RECEIVED is:	*+.43*

This means that the best predictor of FRUIT PRODUCED is FERTILIZER with a correlation of +.77. Adding the other two predictor variables, WATER and LIGHT RECEIVED, improved the prediction by only 0.03 to 0.80, and was, therefore, only slightly better in prediction. FERTILIZER is an excellent predictor of FRUIT PRODUCED by itself.

If you want to learn more about the correlation matrix, see Levine et al. (2011).

7.5 End-of-Chapter Practice Problems

1. Agriculture around the world depends on viable seed production. Crops that originated in various parts of the world are now grown in a wide range of climates. For example, corn (maize) originated in present day Mexico and is now grown throughout the world. The amount of seeds produced in different growing conditions, especially in marginal climates, affect how successful a plant will be for agricultural purposes. The main conditions that can affect plant growth are water, light, and fertilizer. Additionally, "plant crowding" in terms of how close the plants are to each other generally causes negative changes in productivity (i.e., the closer the plants are to one another, the fewer the seeds produced). As a result, agricultural companies have started studying the impact that space between plants has on overall plant production.

   Suppose that you have been asked by a large agricultural seed supplier to analyze some data from their experimental greenhouse. They have asked you to look at how growing conditions predict the amount of seed production for a new type of agricultural plant. You have been asked to use the following growing variables: water (milliliters, ml), light received (in minutes, min), and fertilizer (microliters, μl), and, in addition, to use space (centimeters, cm) between plants as an additional predictor of seed production. The seed supplier has asked you for your recommendation as to whether or not space should be included with the main growing conditions.

You have decided to use a multiple correlation and multiple regression analysis, and to test your Excel skills, you have collected the data from a random sample of seed pods from 12 crop plants that have been grown under controlled conditions during one full growing season. These hypothetical data appear in Fig. 7.10:

SEED PRODUCTION

How well do water, light, fertilizer, and space predict the number of seeds produced per pod?

AVERAGE SEEDS PER POD	WATER (ml)	LIGHT RECEIVED (min)	SPACE (cm)	FERTILIZER (μl)
3.25	600	620	12.5	650
3.42	520	550	10	600
2.85	510	540	5	500
2.65	480	460	2.5	510
3.65	720	710	15	630
3.16	570	610	7.5	550
3.56	710	650	10	610
2.35	500	480	5	430
2.86	450	470	7.5	450
2.95	560	530	10	550
3.15	550	580	10	580
3.45	610	620	12.5	620

Fig. 7.10 Worksheet Data for Chap. 7: Practice Problem #1

(a) Create an Excel spreadsheet using AVERAGE SEEDS PER POD as the criterion (Y), and WATER (X_1), LIGHT RECEIVED (X_2), SPACE (X_3), and FERTILIZER (X_4) as the predictors.

(b) Use Excel's *multiple regression* function to find the relationship between these five variables and place it below the table.

(c) Use number format (two decimal places) for the multiple correlation on the SUMMARY OUTPUT, and use four decimal places for the coefficients in the SUMMARY OUTPUT).

(d) Print the table and regression results below the table so that they fit onto one page.

(e) Save this file as: seed14

Answer the following questions using your Excel printout:

1. What is the multiple correlation R_{xy}?
2. What is the y-intercept a?
3. What is the coefficient for WATER b_1?
4. What is the coefficient for LIGHT RECEIVED b_2?
5. What is the coefficient for SPACE b_3?
6. What is the coefficient for FERTILIZER b_4?
7. What is the multiple regression equation?
8. Predict the AVERAGE SEEDS PER POD you would expect for a WATER score of 610, a LIGHT RECEIVED score of 550, a SPACE score of 3, and a FERTILIZER score of 610.

(f) Now, go back to your Excel file and create a *correlation matrix* for these five variables, and place it underneath the SUMMARY OUTPUT.

(g) Save this file as: seed15

(h) Now, print out *just this correlation matrix* on a separate sheet of paper.

Answer the following questions using your Excel printout. Be sure to include the plus or minus sign for each correlation:

9. What is the correlation between WATER and AVERAGE SEEDS PER POD?

10. What is the correlation between LIGHT RECEIVED and AVERAGE SEEDS PER POD?

11. What is the correlation between SPACE and AVERAGE SEEDS PER POD?

12. What is the correlation between FERTILIZER and AVERAGE SEEDS PER POD?

13. What is the correlation between SPACE and WATER?

14. What is the correlation between LIGHT RECEIVED and FERTILIZER?

15. Discuss which of the four predictors is the best predictor of AVERAGE SEEDS PER POD.

16. Explain in words how much better the four predictor variables together predict AVERAGE SEEDS PER POD than the best single predictor by itself.

2. The Graduate Record Examinations (GRE) are a standardized test that is an admissions requirement for many US graduate schools. The test is intended to measure general academic preparedness, regardless of specialization field. The GRE test produces subtest scores: (1) GRE QUANTITATIVE REASONING (scale 130–170), and (2) WRITING (scale 0–6). Suppose that you have been asked by a top-level MS director of a major graduate program in Environmental Sciences to find the relationship between these variables and FIRST-YEAR GPA based on last year's entering class from the hypothetical data given in Fig. 7.11.

GRADUATE RECORD EXAMINATIONS (GRE)		
FIRST-YEAR GPA	**GRE QUANTITATIVE**	**GRE WRITING**
3.25	160	5
3.42	156	4
2.85	156	2
2.65	154	1
3.65	166	6
3.16	159	3
3.56	166	4
2.35	155	2
2.86	153	3
2.95	158	4
3.15	158	4
3.45	160	5
3.54	165	6
3.36	163	5
3.15	159	4

Fig. 7.11 Worksheet Data for Chap. 7: Practice Problem #2

(a) Create an Excel spreadsheet using FIRST-YEAR GPA as the criterion, and the other variables as the two predictors of this criterion.

(b) Use Excel's **multiple regression** function to find the relationship between these variables and place it below the table.

(c) Use number format (two decimal places) for the multiple correlation on the Summary Output and use two decimal places for the coefficients in the Summary Output.

(d) Print the table and regression results below the table so that they fit onto one page.

(e) By hand on this printout, **circle and label:**

(1a) multiple correlation R
(2b) coefficients for the y-intercept, GRE QUANTITATIVE, GRE WRITING.

(f) On a separate sheet of paper by hand, write the multiple regression equation.

(g) Underneath this regression equation by hand, predict the FIRST-YEAR GPA you would expect for a GRE QUANTITATIVE score of 157, and a GRE WRITING score of 4.

(h) Save this file as: GRE64

(i) Now, go back to your Excel file and create a correlation matrix for these three variables and place it underneath the SUMMARY OUTPUT table. Use two decimal places for all correlations.

(j) Now, print out **just this correlation matrix** on a separate sheet of paper.

(k) By hand underneath the correlation matrix printed out in part j, write the answer to the following questions (label your answer as 1a, 2b, 3c, etc.). Be sure to include the plus or minus sign for each correlation:

(1a) What is the correlation between GRE QUANTITATIVE and FIRST-YEAR GPA?

(2b) What is the correlation between GRE WRITING and FIRST-YEAR GPA?

(3c) What is the correlation between GRE WRITING AND GRE QUANTITATIVE?

(4d) Discuss which predictor is the better single predictor of FIRST-YEAR GPA.

(5e) Explain, in words, how much better the two variables together predict FIRST-YEAR GPA than the better single predictor among the predictor variables.

(6f) Re-save the file as: GRE64

3. Suppose that you have been hired by the United States Department of Agriculture (USDA) to analyze corn yields from Iowa farms over one year (a single growing season). Suppose, further, that these data will represent a pilot study that will be included in a larger ongoing analysis of corn yield in the Midwest. You want to determine if you can predict the amount of corn produced in bushels per acre (bu/acre) based on three predictors: (1) water measured in inches of rainfall per year (in/year), (2) fertilizer measured in the amount of nitrogen applied to the soil in pounds per acre (lbs/acre), and (3) average temperature during the growing season measured in degrees Fahrenheit (°F).

To check your skills in Excel, you have selected a random sample of corn from each of 11 farms selected randomly and recorded the hypothetical data given in Fig. 7.12.

CORN YIELDS ON IOWA FARMS

YIELD (bu/acre)	RAINFALL (in/yr)	NITROGEN (lbs/acre)	TEMPERATURE (degrees F)
240	28.8	190	80
255	29.5	195	79
299	27.6	230	79
230	28.3	175	78
300	31.5	250	79
180	27.1	250	82
294	28.9	220	78
278	27.1	200	74
284	28.8	204	80
194	30.7	185	81
170	27.9	190	88

Fig. 7.12 Worksheet Data for Chap. 7: Practice Problem #3

(a) Create an Excel spreadsheet using YIELD as the criterion (Y), and the other variables as the three predictors of this criterion (X_1 = RAINFALL, X_2 = NITROGEN, and X_3 = TEMPERATURE).

(b) Use Excel's *multiple regression* function to find the relationship between these four variables and place the SUMMARY OUTPUT below the table.

(c) Use number format (two decimal places) for the multiple correlation on the Summary Output and use two decimal places for the coefficients in the SUMMARY OUTPUT.

(d) Save the file as: yield15

(e) Print the table and regression results below the table so that they fit onto one page.

Answer the following questions using your Excel printout:

1. What is the multiple correlation R_{xy}?
2. What is the y-intercept a?
3. What is the coefficient for RAINFALL b_1?
4. What is the coefficient for NITROGEN b_2?
5. What is the coefficient for TEMPERATURE b_3?
6. What is the multiple regression equation?
7. Predict the corn yield you would expect for rainfall of 28 inches per year, nitrogen at 205 pounds/acre, and temperature of 83 degrees Fahrenheit.

(f) Now, go back to your Excel file and create a correlation matrix for these four variables and place it underneath the SUMMARY OUTPUT.

(g) Re-save this file as: yield15

(h) Now, print out *just this correlation matrix* on a separate sheet of paper.

Answer to the following questions using your Excel printout. (Be sure to include the plus or minus sign for each correlation):

8. What is the correlation between RAINFALL and YIELD?
9. What is the correlation between NITROGEN and YIELD?
10. What is the correlation between TEMPERATURE and YIELD?
11. What is the correlation between NITROGEN and RAINFALL?
12. What is the correlation between TEMPERATURE and RAINFALL?
13. What is the correlation between TEMPERATURE and NITROGEN?
14. Discuss which of the three predictors is the best predictor of corn yield.
15. Explain in words how much better the three predictor variables combined predict corn yield than the best single predictor by itself.

References

Hoshmand A R. Statistical Methods for Environmental and Agricultural Sciences (2nd ed.). Boca Raton, FL: CRC Press, 1998.

Keller, G. Statistics for Management and Economics (8th ed.). Mason, OH: South-Western Cengage Learning, 2009.

Levine, D.M., Stephan, D.F., Krehbiel, T.C., and Berenson, M.L. Statistics for Managers using Microsoft Excel (6th ed.). Boston, MA: Prentice Hall/Pearson, 2011.

Chapter 8
One-Way Analysis of Variance (ANOVA)

So far in this 2019 Excel Guide, you have learned how to use a one-group t-test to compare the sample mean to the population mean, and a two-group t-test to test for the difference between two sample means. *But what should you do when you have more than two groups and you want to determine if there is a significant difference between the means of these groups?*

The answer to this question is: *Analysis of Variance (ANOVA).*

The ANOVA test allows you to test for the difference between the means when you have *three or more groups* in your research study.

Important note: In order to do One-way Analysis of Variance, you need to have installed the "Data Analysis Toolpak" that was described in Chap. 6 (see Sect. 6.5.1). If you did not install this, you need to do that now.

Suppose that you were working as a research scientist and that you wanted to do a research study comparing the highway miles per gallon (mpg) for five types of vehicles: (1) SUBCOMPACTS, (2) COMPACTS, (3) MID-SIZE, (4) LARGE, and (5) SUVs. You want to answer the research question: Is the size of the vehicle related to gasoline usage? You have obtained the cooperation of the owners of each type of vehicle who agree to keep track of their highway mileage over a predetermined route for three tanks of gasoline. The hypothetical data for this study are given in Fig. 8.1.

T. J. Quirk et al., *Excel 2019 for Environmental Sciences Statistics*,
Excel for Statistics, https://doi.org/10.1007/978-3-030-66277-6_8

HIGHWAY MILES PER GALLON (mpg) COMPARISON OF FIVE TYPES OF CARS

1	2	3	4	5
SUBCOMPACTS (mpg)	COMPACTS (mpg)	MID-SIZE (mpg)	LARGE (mpg)	SUVs (mpg)
28.1	26.2	24.0	22.0	18.1
30.2	28.3	26.3	23.1	20.2
29.3	29.3	25.2	25.4	22.3
31.6	27.0	27.1	24.3	21.4
33.0	28.0	28.0	25.0	20.5
34.3	29.5	23.6	24.7	19.0
32.1	31.0	29.2	23.1	18.2
35.0	32.3		22.4	19.1
	33.1		26.0	
			21.3	

Fig. 8.1 Worksheet Data for Highway mpg Test

Note that each type of car can have a different number of vehicles in it in order for ANOVA to be used on the data. Statisticians delight in this fact by referring to this characteristic by stating that: "ANOVA is a very robust test." (Statisticians love that term!)

Create an Excel spreadsheet for these data in this way:

A4:   HIGHWAY MILES PER GALLON (mpg) COMPARISON OF FIVE TYPES OF CARS
B7:   SUBCOMPACTS (mpg)
B8:   28.1
B15:   35.0
C7:   COMPACTS (mpg)
D7:   MID-SIZE (mpg)
E7:   LARGE (mpg)
E17:   21.3
F7:   SUVs (mpg)
F8:   18.1
F15:   19.1

Enter the other information into your spreadsheet table. When you have finished entering these data, the last cell on the left should have 35.0 in cell B15, and the last cell on the right should have 19.1 in cell F15. Center the numbers and labels in each of the columns so that the data look like Fig. 8.1. Use number format (one decimal) for all numbers.

Important note: Be sure to double-check all of your figures in the table to make sure that they are exactly correct or you will not be able to obtain the correct answer for this problem!

Save this file as: CARS2

8.1   Using Excel to Perform a One-Way Analysis of Variance (ANOVA)

Objective: To use Excel to perform a one-way ANOVA test.

You are now ready to perform an ANOVA test on these data using the following steps:

Data (at top of screen)

Data Analysis (far right at top of screen)

Anova: Single Factor (*scroll up to this formula and highlight it; see Fig. 8.2*)

Fig. 8.2 Dialog Box for Data Analysis: Anova Single Factor

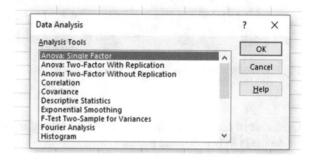

OK

Input range: B7:F17 (note that you have included in this range the column titles that are in row 7)

Important note: When you define the Input Range of the data, be sure that it includes only the data that you are measuring.

Important note: Whenever the data set has a different sample size in the groups being compared, the INPUT RANGE that you define must start at the column title of the first group on the left and go to the last column on the right to the lowest row that has a figure in it in the entire data matrix so that the INPUT RANGE has the "shape" of a rectangle when you highlight it. Since LARGE has 21.3 in cell E17, your "rectangle" must include row 17!

Grouped by: Columns

Put a check mark in: Labels in First Row

Output range (click on the button to its left): A19 (see Fig. 8.3)

Fig. 8.3 Dialog Box for Anova: Single Factor Input/Output Range

OK

Center all of the numbers in the ANOVA table, and round off all numbers that are decimals to two decimal places.

Save this file as: CARS2A

You should have generated the table given in Fig. 8.4.

HIGHWAY MILES PER GALLON (mpg) COMPARISON OF FIVE TYPES OF CARS

	1	2	3	4	5
	SUBCOMPACTS (mpg)	COMPACTS (mpg)	MID-SIZE (mpg)	LARGE (mpg)	SUVs (mpg)
	28.1	26.2	24.0	22.0	18.1
	30.2	28.3	26.3	23.1	20.2
	29.3	29.3	25.2	25.4	22.3
	31.6	27.0	27.1	24.3	21.4
	33.0	28.0	28.0	25.0	20.5
	34.3	29.5	23.6	24.7	19.0
	32.1	31.0	29.2	23.1	18.2
	35.0	32.3		22.4	19.1
		33.1		26.0	
				21.3	

Anova: Single Factor

SUMMARY

Groups	Count	Sum	Average	Variance
SUBCOMPACTS (mpg)	8	253.6	31.70	5.78
COMPACTS (mpg)	9	264.7	29.41	5.48
MID-SIZE (mpg)	7	183.4	26.20	4.28
LARGE (mpg)	10	237.3	23.73	2.48
SUVs (mpg)	8	158.8	19.85	2.29

ANOVA

Source of Variation	SS	df	MS	F	P-value	F crit
Between Groups	718.23	4	179.56	44.80	1.07093E-13	2.63
Within Groups	148.29	37	4.01			
Total	866.52	41				

Fig. 8.4 ANOVA Results for Highway mpg Test

Print out both the data table and the ANOVA summary table so that all of this information fits onto one page. (Hint: Set the Page Layout/Fit to Scale to *75% size*).

As a check on your analysis, you should have the following in these cells:

A19:   Anova: Single Factor
D24:   29.41
E32:   44.80
G32:   2.63
C35:   41

Re-save the file as: CARS2A

Now, let's discuss how you should interpret this table:

8.2 How to Interpret the ANOVA Table Correctly

Objective: To interpret the ANOVA table correctly

ANOVA allows you to test for the differences between means when you have three or more groups of data. This ANOVA test is called the F-test statistic and is typically identified with the letter: F.

The formula for the F-test is this:

F = Mean Square between groups (MS_b) divided by Mean Square within groups (MS_w)

$$F = MS_b/MS_w \tag{8.1}$$

The derivation and explanation of this formula is beyond the scope of this *Excel Guide*. In this *Excel Guide*, we are attempting to teach you *how to use Excel*, and we are not attempting to teach you the statistical theory that is behind the ANOVA formulas. For a detailed explanation of ANOVA, see Hibbert and Gooding (2006) and Hochmand (1998).

Note that cell D32 contains $MS_b = 179.56$, while cell D33 contains $MS_w = 4.01$.

When you divide these two figures using their cell references in Excel, you get the answer for the F-test of 44.80 which is in cell E32. (Remember, Excel is more accurate than your calculator!) Let's discuss now the meaning of the figure: $F = 44.80$.

In order to determine whether this figure for F of 44.80 indicates a significant difference between the means of the groups, the first step is to write the null hypothesis and the research hypothesis for the five types of cars.

In our statistics highway mpg comparisons, the null hypothesis states that the population means of the five groups are equal, while the research hypothesis states that the population means of the five groups are not equal and that there is, therefore, a significant difference between the population means of the five groups. Which of these two hypotheses should you accept based on the ANOVA results?

8.3 Using the Decision Rule for the ANOVA F-Test

To state the hypotheses, let's call SUBCOMPACTS as Group 1, COMPACTS as Group 2, MID-SIZE as Group 3, LARGE as Group 4, and SUVs as Group 5. The hypotheses would then be:

$H_0 : \mu_1 = \mu_2 = \mu_3 = \mu_4 = \mu_5$

$H_1 : \mu_1 \neq \mu_2 \neq \mu_3 \neq \mu_4 \neq \mu_5$

The answer to this question is analogous to the decision rule used in this book for both the one-group t-test and the two-group t-test. You will recall that this rule (See Sects. 4.1.6 and 5.1.8) was:

If the absolute value of t is less than the critical t, you accept the null hypothesis.

or

If the absolute value of t is greater than the critical t, you reject the null hypothesis and accept the research hypothesis.

Now, here is the decision rule for ANOVA:

Objective: To learn the decision rule for the ANOVA F-test

The decision rule for the ANOVA F-test is the following:

If the value for F is less than the critical F-value, accept the null hypothesis.

or

If the value of F is greater than the critical F-value, reject the null hypothesis and accept the research hypothesis.

Note that Excel tells you the critical F-value in cell G32: 2.63

Therefore, our decision rule for the types of cars ANOVA test is this:

Since the value of F of 44.80 is greater than the critical F-value of 2.63, we reject the null hypothesis and accept the research hypothesis.

Therefore, our conclusion, in plain English, is:

There is a significant difference between the highway mpg between the five types of cars.

Note that it is not necessary to take the absolute value of F of 44.80. The F-value can never be less than one, and so it can never be a negative value which requires us to take its absolute value in order to treat it as a positive value.

It is important to note that ANOVA tells us that there was a significant difference between the population means of the groups, *but it does not tell us which pairs of groups were significantly different from each other.*

8.4 Testing the Difference Between Two Groups Using the ANOVA t-Test

To answer that question, we need to do a different test called the ANOVA t-test.

Objective: To test the difference between the means of two groups using an ANOVA t-test when the ANOVA F-test results indicate a significant difference between the population means.

Since we have five groups of data (one group for each of five types of cars), we would have to perform ten separate ANOVA t-tests to determine which of the ten pairs of groups were significantly different. This requires that we would have to perform a separate ANOVA t-test for the following ten pairs of groups:

 (1) SUBCOMPACTS vs. COMPACTS
 (2) SUBCOMPACTS vs. MID-SIZE
 (3) SUBCOMPACTS vs. LARGE
 (4) SUBCOMPACTS vs. SUVs
 (5) COMPACTS vs. MID-SIZE
 (6) COMPACTS vs. LARGE
 (7) COMPACTS vs. SUVs
 (8) MID-SIZE vs. LARGE
 (9) MID-SIZE vs. SUVs
(10) LARGE vs. SUVs

We will do just one of these pairs of tests, COMPACTS vs. LARGE, to illustrate the way to perform an ANOVA t-test comparing these two types of cars. The ANOVA t-test for the other nine pairs of groups would be done in the same way.

8.4.1   Comparing COMPACTS vs. LARGE in Highway mpg Using the ANOVA t-Test

> Objective: To compare COMPACTS vs. LARGE in highway mpg using the
> ANOVA t-test

The first step is to write the null hypothesis and the research hypothesis for these two types of cars.

For the ANOVA t-test, the null hypothesis is that the population means of the two groups are equal, while the research hypothesis is that the population means of the two groups are not equal (i.e., there is a significant difference between these two means). Since we are comparing COMPACTS (Group 2) vs. LARGE (Group 4), these hypotheses would be:

$H_0 : \quad \mu_2 = \mu_4$

$H_1 : \quad \mu_2 \neq \mu_4$

For Group 2 vs. Group 4, the formula for the ANOVA t-test is:

$$ANOVA\ t = \frac{\overline{X}_1 - \overline{X}_2}{s.e._{ANOVA}} \tag{8.2}$$

where

$$s.e._{ANOVA} = \sqrt{MS_w \left(\frac{1}{n_1} + \frac{1}{n_2} \right)} \tag{8.3}$$

The steps involved in computing this ANOVA t-test are:

1. Find the difference of the sample means for the two groups ($29.41 - 23.73 = 5.68$).
2. Find $1/n_2 + 1/n_4$ (since both groups have a different number of cars in them, this becomes: $1/9 + 1/10 = 0.11 + 0.10 = 0.21$).
3. Multiply MS_w times the answer for step 2 ($4.01 \times 0.21 = 0.84$)
4. Take the square root of step 3 (SQRT $(0.84) = 0.92$)
5. Divide Step 1 by Step 4 to find ANOVA t ($5.68/0.92 = 6.17$)

Note: Since Excel computes all calculations to 16 decimal places, when you use Excel for the above computations, your answer will be 6.18 in two decimal places, but Excel's answer will be much more accurate because it is always in 16 decimal places in its computations.

Now, what do we do with this ANOVA t-test result of 6.18? In order to interpret this value of 6.18 correctly, we need to determine the critical value of t for the ANOVA t-test. To do that, we need to find the degrees of freedom for the ANOVA t-test as follows:

8.4.1.1 Finding the Degrees of Freedom for the ANOVA t-Test

Objective: To find the degrees of freedom for the ANOVA t-test

The degrees of freedom (df) for the ANOVA t-test is found as follows:

df = take *the total sample size of all of the groups* and subtract the number of groups in your study ($n_{TOTAL} - k$ where k = the number of groups)

In our example, the total sample size of the five groups is 42 since there are 8 cars in Group 1, 9 cars in Group 2, 7 cars in Group 3, 10 cars in Group 4, and 8 cars in Group 5, and since there are five groups, $42 - 5$ gives a degrees of freedom for the ANOVA t-test of 37.

If you look up df = 37 in the **t-table in Appendix E in the degrees of freedom column (df), which is the second column on the left of this table**, you will find that the critical t-value is 2.026.

Important note: Be sure to use the degrees of freedom column (df) in Appendix E for the ANOVA t-test critical t-value

8.4.1.2 Stating the Decision Rule for the ANOVA t-Test

> Objective: To learn the decision rule for the ANOVA t-test

Interpreting the result of the ANOVA t-test follows the same decision rule that we used for both the one-group t-test (see Sect. 4.1.6) and the two-group t-test (see Sect. 5.1.8):

If the absolute value of t is less than the critical value of t, we accept the null hypothesis.

or

If the absolute value of t is greater than the critical value of t, we reject the null hypothesis and accept the research hypothesis.

Since we are using a type of t-test, we need to take the absolute value of t. Since the absolute value of 6.18 is greater than the critical t-value of 2.026, we reject the null hypothesis (that the population means of the two groups are equal) and accept the research hypothesis (that the population means of the two groups are significantly different from one another).

This means that our conclusion, in plain English, is as follows:

The average highway mpg for COMPACTS was significantly higher than the average for LARGE (29 vs. 24).

8.4.1.3 Performing an ANOVA t-Test Using Excel Commands

Now, let's do these calculations for the ANOVA t-test using Excel with the file you created earlier in this chapter: CARS2A

A37: COMPACTS vs. LARGE
A39: 1/9 + 1/10
A41: s.e. ANOVA
A43: ANOVA t-test
B39: =(1/9 + 1/10)   (no spaces between)
B41: =SQRT(D33*B39)   (no spaces between)
B43: =(D24 − D26)/B41   (no spaces between)

You should now have the following results in these cells when you round off all these figures in the ANOVA t-test to two decimal points.:

B39: 0.21
B41: 0.92
B43: 6.18

Save this final result under the file name: CARS3

Print out the resulting spreadsheet so that it fits onto one page like Fig. 8.5 (Hint: Reduce the Page Layout/Scale to Fit to 75%).

HIGHWAY MILES PER GALLON (mpg) COMPARISON OF FIVE TYPES OF CARS

	1	2	3	4	5
	SUBCOMPACTS (mpg)	COMPACTS (mpg)	MID-SIZE (mpg)	LARGE (mpg)	SUVs (mpg)
	28.1	26.2	24.0	22.0	18.1
	30.2	28.3	26.3	23.1	20.2
	29.3	29.3	25.2	25.4	22.3
	31.6	27.0	27.1	24.3	21.4
	33.0	28.0	28.0	25.0	20.5
	34.3	29.5	23.6	24.7	19.0
	32.1	31.0	29.2	23.1	18.2
	35.0	32.3		22.4	19.1
		33.1		26.0	
				21.3	

Anova: Single Factor

SUMMARY

Groups	Count	Sum	Average	Variance
SUBCOMPACTS	8	253.60	31.70	5.78
COMPACTS	9	264.70	29.41	5.48
MID-SIZE	7	183.40	26.20	4.28
LARGE	10	237.30	23.73	2.48
SUVs	8	158.80	19.85	2.29

ANOVA

Source of Variation	SS	df	MS	F	P-value	F crit
Between Groups	718.23	4	179.56	44.80	1.07093E-13	2.63
Within Groups	148.29	37	4.01			
Total	866.52	41				

COMPACTS vs. LARGE

1/9+1/10	0.21

s.e. ANOVA	0.92

ANOVA t-test	6.18

Fig. 8.5 Final Spreadsheet of Highway mpg for COMPACTS vs. LARGE

For a more detailed explanation of the ANOVA t-test, see Townend (2002).

Important note: You are only allowed to perform an ANOVA t-test comparing the means of two groups when the F-test produces a significant difference between the means of all of the groups in your study.

It is improper to do any ANOVA t-test when the value of F is less than the critical value of F. Whenever F is less than the critical F, this

means that there was no difference between the means of the groups, and, therefore, that you cannot test to see if there is a difference between the means of any two groups since this would capitalize on chance differences between these two groups. For more information on this important point, see Gould and Gould (2002).

8.5   End-of-Chapter Practice Problems

1. Let's suppose that you have been asked to study the yield (grams of product produced) of a chemical reaction conducted under three different temperature conditions: (1) BELOW ROOM TEMPERATURE (15 degrees Celsius (°C)), (2) ROOM TEMPERATURE (25 degrees Celsius (°C)), and (3) ABOVE ROOM TEMPERATURE (30 degrees Celsius (°C)).

   You have been asked to analyze the data from the reactions to determine if there was a significant difference in yield (grams of product produced) between the three different temperatures. To test your Excel skills, you have selected a random sample of results from each of the three chemical reactions performed at different temperatures (see Fig. 8.6). Note that each group reactions can be of a different number of results in order for ANOVA to be used on the data. Statisticians delight in this fact by referring to this characteristic by stating that: "ANOVA is a very robust test." (Statisticians love that term!)

CHEMICAL REACTION YIELD (grams of product produced)

BELOW ROOM TEMP (15 °C)	ROOM TEMP (25 °C)	ABOVE ROOM TEMP (30 °C)
90	85	76
85	89	80
74	83	90
89	79	84
84	74	78
95	75	65
92	86	42
65	87	58
75	86	63
73	88	75
54		66
71		

Fig. 8.6 Worksheet Data for Chap. 8: Practice Problem #1

(a) Enter these data on an Excel spreadsheet.
(b) Perform a *one-way ANOVA test* on these data and show the resulting ANOVA table *underneath* the input data for the three temperatures.
(c) If the F-value in the ANOVA table is significant, create an Excel formula to compute the ANOVA t-test comparing the average for ROOM TEMPERATURE against ABOVE ROOM TEMPERATURE and show the results below the ANOVA table on the spreadsheet (put the standard error and the ANOVA t-test value on separate lines of your spreadsheet and use two decimal places for each value)
(d) Print out the resulting spreadsheet so that all of the information fits onto one page.
(e) Save the spreadsheet as: REACTION3

Now, write the answers to the following questions using your Excel printout:

1. What are the null hypothesis and the research hypothesis for the ANOVA F-test?
2. What is MS_b on your Excel printout?
3. What is MS_w on your Excel printout?
4. Compute $F = MS_b/MS_w$ using your calculator.
5. What is the critical value of F on your Excel printout?
6. What is the result of the ANOVA F-test?
7. What is the conclusion of the ANOVA F-test in plain English?
8. If the ANOVA F-test produced a significant difference between the three types of temperatures, what is the null hypothesis and the research hypothesis for the ANOVA t-test comparing ROOM TEMPERATURE versus ABOVE ROOM TEMPERATURE?
9. What is the mean (average) for ROOM TEMPERATURE on your Excel printout?
10. What is the mean (average) for ABOVE ROOM TEMPERATURE on your Excel printout?
11. What are the degrees of freedom (df) for the ANOVA t-test comparing ROOM TEMPERATURE versus ABOVE ROOM TEMPERATURE?
12. What is the critical t-value for this ANOVA t-test in Appendix E for these degrees of freedom?
13. Compute the $s.e._{ANOVA}$ using your calculator.
14. Compute the ANOVA t-test value comparing ROOM TEMPERATURE versus ABOVE ROOM TEMPERATURE using your calculator.
15. What is the result of the ANOVA t-test comparing ROOM TEMPERATURE versus ABOVE ROOM TEMPERATURE?
16. What is the conclusion of the ANOVA t-test comparing ROOM TEMPERATURE versus ABOVE ROOM TEMPERATURE in plain English?

Note that since there are three types of temperatures, you need to do three ANOVA t-tests to determine what the significant differences are between the three types of temperatures. *Since you have just completed the ANOVA t-test comparing   ROOM   TEMPERATURE   versus   ABOVE   ROOM*

TEMPERATURE, *you would also need to do the ANOVA t-test comparing* ROOM TEMPERATURE *versus* BELOW ROOM TEMPERATURE, *and also the ANOVA t-test comparing* ABOVE ROOM TEMPERATURE *versus* BELOW ROOM TEMPERATURE *in order to write a conclusion summarizing these three types of ANOVA t-tests.*

2. In a controlled greenhouse experiment, what type of soil conditions are favorable for the growth of a common flowering species such as Daisy Fleabane (*Erigeron annuus*) often found along roadsides? Suppose that you wanted to conduct a controlled research study to compare the growth of this flower in three types of soil treatments, specifically the effects of Nitrogen. Nitrogen is an essential macro-nutrient necessary for plant development and is commonly found in fertilizers: (1) Group 1: No nitrogen added to the soil, (2) Group 2: Low amounts of Nitrogen added to the soil, and (3) Group 3: High amounts of Nitrogen added to the soil. Your measured variable will be milligrams (mg) of dry biomass of *Erigeron annuus*. To do this study, you have obtained soil from one location of land in the state of Colorado and seeds from a distributor. You have sterilized the soil so that you could place it in 12 containers for each type of treatment (i.e., 36 containers in all). Suppose, further, that you planted the same number of seeds of *Erigeron annuus* in each container and maintained the containers in a controlled greenhouse under constant conditions for a growing season. You then clipped and dried all the *Erigeron annuus* from each treatment to obtain dry biomass. You then weighed the dry biomass of each treatment and generated the hypothetical data given in Fig. 8.7. Note that there are a different number of containers in each of the three treatments because some of the plants did not survive the growing season. To test your Excel skills, you have decided to run an ANOVA test of the data to determine the growing characteristics of this plant with respect to varying amounts of Nitrogen.

DAISY FLEABANE GROWTH IN ELEVATED NITROGEN SOILS

No Nitrogen	Low Nitrogen	High Nitrogen
500	550	800
550	600	1000
700	750	900
650	700	1100
500	600	1400
550	650	1600
600	600	1500
650	700	1800
600	650	1500
	750	1800
		2000
		1600

Fig. 8.7 Worksheet Data for Chap. 8: Practice Problem #2

(a) Enter these data on an Excel spreadsheet.

(b) Perform a *one-way ANOVA test* on these data and show the resulting ANOVA table *underneath* the input data for the three levels of Nitrogen.

(c) If the F-value in the ANOVA table is significant, create an Excel formula to compute the ANOVA t-test comparing the average for Low Nitrogen against High Nitrogen and show the results below the ANOVA table on the spreadsheet (put the standard error and the ANOVA t-test value on separate lines of your spreadsheet and use two decimal places for each value).

(d) Print out the resulting spreadsheet so that all of the information fits onto one page.

(e) Save the spreadsheet as: Weed5

Now, write the answers to the following questions using your Excel printout:

1. What are the null hypothesis and the research hypothesis for the ANOVA F-test?
2. What is MS_b on your Excel printout?
3. What is MS_w on your Excel printout?
4. Compute $F = MS_b/MS_w$ using your calculator.
5. What is the critical value of F on your Excel printout?
6. What is the result of the ANOVA F-test?
7. What is the conclusion of the ANOVA F-test in plain English?
8. If the ANOVA F-test produced a significant difference between the three amounts of Nitrogen in the weight of the flowers grown in the greenhouse during a growing season, what is the null hypothesis and the research hypothesis for the ANOVA t-test comparing Low Nitrogen versus High Nitrogen?
9. What is the mean (average) for Low Nitrogen on your Excel printout?
10. What is the mean (average) for High Nitrogen on your Excel printout?
11. What are the degrees of freedom (df) for the ANOVA t-test comparing Low Nitrogen versus High Nitrogen?
12. What is the critical t-value for this ANOVA t-test in Appendix E for these degrees of freedom?
13. Compute the s.e.$_{ANOVA}$ using your calculator.
14. Compute the ANOVA t-test value comparing Low Nitrogen versus High Nitrogen using your calculator.
15. What is the result of the ANOVA t-test comparing Low Nitrogen versus High Nitrogen?
16. What is the conclusion of the ANOVA t-test comparing Low Nitrogen versus High Nitrogen in plain English?

Note that since there are three treatments, you need to do three ANOVA t-tests to determine what the significant differences are between the three treatments. *Since you have just completed the ANOVA t-test comparing Low Nitrogen versus High Nitrogen, you would also need to do the ANOVA t-test comparing Low Nitrogen versus No Nitrogen, and also the ANOVA t-test comparing No Nitrogen versus High Nitrogen in order to write a conclusion summarizing all of these three types of ANOVA t-tests.*

3. Rivers supply drinking water, water for irrigation, water for sewage disposal, and for many other functions. Suppose that you have been hired as a research assistant to "run the data" from a study of a river in which water samples were taken at three locations: (1) LOCATION A: One mile upstream from an industrial plant located on the shore of the river, (2) LOCATION B: One mile downstream from an industrial plant located on the shore of the river, and (3) LOCATION C: two miles outside a city where treated sewage is released into the river. Water samples were taken randomly from the river and were measured in parts per million (ppm) of dissolved oxygen content. To test your Excel skills, you have created the hypothetical data in Fig. 8.8

Fig. 8.8 Worksheet Data
for Chap. 8: Practice
Problem #3

DISSOLVED OXYGEN CONTENT IN WATER (ppm)

	LOCATION	
A	**B**	**C**
6.2	3.8	3.9
6.9	4.9	4.9
6.8	3.9	4.1
6.6	4.1	4.3
6.5	4.6	4.6
6.4	4.2	4.5
6.9	4.6	4.7
6.7	4.7	4.8
6.5	4.8	4.7
6.8	4.4	4.6

(a) Enter these data on an Excel spreadsheet.
(b) Perform a *one-way ANOVA test* on these data and show the resulting ANOVA table *underneath* the input data for the three locations.
(c) If the F-value in the ANOVA table is significant, create an Excel formula to compute the ANOVA t-test comparing the average ppm for LOCATION A against the average ppm for LOCATION B and show the results below the ANOVA table on the spreadsheet (put the standard error and the ANOVA t-test value on separate lines of your spreadsheet and use two decimal places for each value)
(d) Print out the resulting spreadsheet so that all of the information fits onto one page.
(e) Save the spreadsheet as: WATER14

Now, write the answers to the following questions using your Excel printout:

1. What are the null hypothesis and the research hypothesis for the ANOVA F-test?
2. What is MS_b on your Excel printout?
3. What is MS_w on your Excel printout?
4. Compute $F = MS_b/MS_w$ using Excel.
5. What is the critical value of F on your Excel printout?
6. What is the result of the ANOVA F-test?
7. What is the conclusion of the ANOVA F-test in plain English?
8. If the ANOVA F-test produced a significant difference in the average ppm between the three locations, what is the null hypothesis and the research hypothesis for the ANOVA t-test comparing LOCATION A versus LOCATION B?
9. What is the mean (average) ppm for LOCATION A on your Excel printout?
10. What is the mean (average) ppm for LOCATION B on your Excel printout?
11. What are the degrees of freedom (df) for the ANOVA t-test comparing LOCATION A vs. LOCATION B?
12. What is the critical t-value for this ANOVA t-test in Appendix E for these degrees of freedom?
13. Compute the s.e.$_{ANOVA}$ using Excel.
14. Compute the ANOVA t-test value comparing LOCATION A vs. LOCATION B using Excel.
15. What is the result of the ANOVA t-test comparing LOCATION A vs. LOCATION B?
16. What is the conclusion of the ANOVA t-test comparing LOCATION A vs. LOCATION B in plain English?

References

Gould J, Gould G. Biostats basics: a student handbook. New York: W.H. Freeman and Company; 2002.

Hibbert D, Gooding J. Data analysis for chemistry: an introductory guide for students and laboratory scientists. New York: Oxford University Press; 2006.

Hochmand A R. Statistical methods for environmental and agricultural sciences (2nd ed.). Boca Raton, FL: CRC Press; 1998.

Townend J. Practical statistics for environmental and biological scientists. Hoboken, NJ: John Wiley & Sons; 2002.

Appendices

Appendix A: Answers to End-of-Chapter Practice Problems

T. J. Quirk et al., *Excel 2019 for Environmental Sciences Statistics*,
Excel for Statistics, https://doi.org/10.1007/978-3-030-66277-6

Chapter 1: Practice Problem #1 Answer (see Fig. A.1)

CHLORINE RESIDUAL IN A WASTE TREATMENT PLANT		
DATA (mg/L)		
0.9		
1.1	n	17
1.3		
1.0		
0.8	Mean	1.93
0.9		
1.0		
1.1	STDEV	2.48
1.3		
1.2		
8.0	s.e.	0.60
9.0		
1.2		
1.0		
1.3		
0.8		
0.9		

Fig. A.1 Answer to Chap. 1: Practice Problem #1

Chapter 1: Practice Problem #2 Answer (see Fig. A.2)

Chapter 1: Practice Problem #2 answer (see Fig.A.2)

LEAD CONCENTRATION IN AIR SAMPLES TAKEN NEAR SAN FRANCISCO

Micrograms per cubic meter ($\mu g/m^3$)		
3.1		
10.1	n	14
6.7		
8.9		
5.6	Mean	7.44
6.4		
4.8		
10.2	STDEV	2.25
9.8		
8.4		
7.5	s.e.	0.60
9.4		
8.5		
4.8		

Fig. A.2 Answer to Chap. 1: Practice Problem #2

Chapter 1: Practice Problem #3 Answer (see Fig. A.3)

Chapter 1: Practice problem #3 answer (see Fig. A.3)

MEASUREMENTS OF TcCB IN AN UNCONTAMINATED SITE

parts per billion (ppb)		
0.25		
0.26	n	12
0.28		
0.24		
1.11	MEAN	0.658
1.31		
1.21		
0.72	STDEV	0.406
0.84		
0.28		
0.53	s.e.	0.117
0.86		

Fig. A.3 Answer to Chap. 1: Practice Problem #3

Chapter 2: Practice Problem #1 Answer (see Fig. A.4)

Chapter 2: Practice problem #1 answer (see Fig. A.4)

FRAME NUMBERS	Duplicate frame numbers	RAND NO.
1	8	0.764
2	22	0.913
3	31	0.578
4	42	0.005
5	4	0.407
6	29	0.867
7	3	0.786
8	21	0.353
9	37	0.875
10	17	0.184
11	34	0.633
12	25	0.083
13	10	0.492
14	41	0.645
15	30	0.842
16	36	0.049
17	13	0.892
18	15	0.933
19	20	0.703
20	14	0.741
21	9	0.293
22	12	0.440
23	38	0.475
24	26	0.636
25	1	0.903
26	5	0.158
27	35	0.827
28	28	0.910
29	24	0.364
30	32	0.871
31	27	0.502
32	19	0.867
33	6	0.535
34	39	0.907
35	2	0.576
36	18	0.932
37	7	0.548
38	11	0.027
39	16	0.248
40	40	0.859
41	33	0.234
42	23	0.637

Fig. A.4 Answer to Chap. 2: Practice Problem #1

Chapter 2: Practice Problem #2 Answer (see Fig. A.5)

Fig. A.5 Answer to
Chap. 2: Practice Problem #2

Chapter 2: Practice problem #2 answer (see Fig.A.5)

FRAME NO.	Duplicate frame numbers	Random number
1	50	0.960
2	42	0.188
3	82	0.156
4	59	0.389
5	74	0.925
6	34	0.518
7	67	0.162
8	29	0.821
9	68	0.174
10	71	0.400
11	1	0.184
12	14	0.862
13	10	0.388
14	47	0.423
15	8	0.191
16	69	0.162
17	12	0.486
18	81	0.626
19	7	0.510
20	38	0.633
21	21	0.458
22	57	0.112
23	77	0.650
24	41	0.530
25	23	0.331
26	32	0.227
27	76	0.301
28	60	0.253
29	18	0.541
30	40	0.937
31	4	32
70		0.50.
71	72	0.063
72	54	0.520
73	26	0.877
74	24	0.732
75	85	0.218
76	45	0.613
77	73	0.775
78	43	0.189
79	84	0.290
80	11	0.932
81	30	0.500
82	65	0.297
83	61	0.140
84	64	0.688
85	33	0.489
86	62	0.128

Chapter 2: Practice Problem #3 Answer (see Fig. A.6)

Fig. A.6 Answer to
Chap. 2: Practice Problem #3

Chapter 2: Practice problem #3 answer (see Fig. A.6)

FRAME NUMBERS	Duplicate frame numbers	Random number
1	58	0.350
2	7	0.705
3	37	0.932
4	49	0.496
5	26	0.159
6	65	0.114
7	48	0.313
8	63	0.505
9	15	0.550
10	25	0.441
11	21	0.419
12	36	0.544
13	43	0.183
14	11	0.576
15	10	0.793
16	39	0.502
17	72	0.947
18	59	0.331
19	16	0.453
20	54	0.717
21	52	0.131
22	3	0.098
23	45	0.358
24	4	0.279
25	76	0.246
26		

	33	0.0.
60	31	0.926
61	35	0.926
62	51	0.634
63	62	0.996
64	41	0.294
65	47	0.520
66	14	0.035
67	13	0.944
68	38	0.727
69	17	0.746
70	57	0.571
71	60	0.997
72	9	0.178
73	20	0.293
74	18	0.090
75	5	0.502

Chapter 3: Practice Problem #1 Answer (see Fig. A.7)

Chapter 3: Practice problem #1 answer (see Fig. A.7)

SMALL LAKES IN MICHIGAN

SULFATE LEVELS (SO$_4$ in mg/L)				
1.5				
1.7	Null hypothesis:	μ	=	4.65 mg/L
5.5				
4.3	Research hypothesis:	μ	$\neq$	4.65 mg/L
3.9				
7.4				
1.5	n		19	
2.7				
3.3	Mean		4.06	
7.1				
6.8	STDEV		2.15	
6.3				
5.7	s.e.		0.49	
5.4				
6.2				
1.6	95% confidence interval			
1.8				
1.9		lower limit	3.02	
2.5				
		upper limit	5.10	

	3.02 ------------------- 4.06------- ----4.65--- -------5.10--- ------
	lower	mean	Ref.	upper
	limit		Value	limit

Result: Since the reference value is inside the confidence interval, we accept H$_0$

Conclusion: The was no difference in the sulfate levels in the Michigan lakes this year compared to five years ago.

Fig. A.7 Answer to Chap. 3: Practice Problem #1

Chapter 3: Practice Problem #2 Answer (see Fig. A.8)

WEED KILLER INVENTORY FOR BRAND A

Question: "Is an inventory level of 12 gallons per business day sufficient for Brand A?"

DATA (gallons/day)			
11			
13	Null hypothesis:	μ = 12 gal/business day	
12			
10	Research hypothesis:	μ ≠ 12 gal/business day	
11			
13			
14	n	16	
13			
12			
14	Mean	12.38	
13			
13			
13	STDEV	1.31	
12			
14			
10	s.e.	0.33	

95% confidence interval

	lower limit	11.68
	upper limit	13.07

--------11.68-----------	------12--------	---12.38--	-------------- 13.07------
lower	Ref.	Mean	upper
limit	Value		limit

Result: Since the reference value is inside the confidence interval, we accept the null hypothesis.

Conclusion: An inventory level of 12 gallons sold per business day is sufficient for Brand A weed killer.

Fig. A.8 Answer to Chap. 3: Practice Problem #2

Chapter 3: Practice Problem #3 Answer (see Fig. A.9)

Chapter 3: Practice problem #3 answer (see Fig. A.9)

CONCENTRATION OF SO_2 IN THE ATMOSPHERE IN PARTS PER BILLION (ppb)

ppb					
390	Null hypothesis:	μ	$=$	120 ppb	
332					
186	Research hypothesis:	μ	$\neq$	120 ppb	
85					
29					
135	n		23		
86					
54	mean		73.13		
28					
35	stdev		100.25		
37					
28	s.e.		20.90		
18					
32					
24	95% confidence interval				
19					
21		lower limit		29.78	
35					
31		upper limit		116.48	
18					
20		29.78 ----------- 73.13----- ----------- 116.48 ----------- 120 -------			
21		lower	mean	upper	Ref.
18		limit		limit	Value

Result: Since the reference value is outside the confidence interval, we reject
the null hypothesis and accept the research hypothesis.

Conclusion: The concentration of SO_2 in the atmosphere in the selected sites was
significantly less this year than it was three years ago, and it is now
probably closer to 73 parts per billion.

Fig. A.9   Answer to Chap. 3: Practice Problem #3

Chapter 4: Practice Problem #1 Answer (see Fig. A.10)

Chapter 4: Practice problem #1 answer (see Fig. A.10)

PHOSPORUS CONCENTRATION (mg/L) IN WASTE WATER EFFLUENT

CONCENTRATION (mg/L)				
0.0142	Null hypothesis:	μ	=	0.015 mg/L
0.0135				
0.0138	Research hypothesis:	μ	$\neq$	0.015 mg/L
0.0136				
0.0137	n	15		
0.0135				
0.0141				
0.0140	Mean	0.0137		
0.0138				
0.0134				
0.0135	STDEV	0.0003		
0.0137				
0.0142				
0.0132	s.e.	0.0001		
0.0133				
	critical t	2.145		
	t-test	-15.92		

Result: Since the absolute value of – 15.92 is greater than the critical t of 2.145, we reject the null hypothesis and accept the research hypothesis.

Conclusion: There was significantly less total phosporus concentration in the waste water effluent produced by the chemical plant than the chemical standard of 0.015 mg/L, and it was probably closer tp 0.0137 mg/L.

Fig. A.10 Answer to Chap. 4: Practice Problem #1

Chapter 4: Practice Problem #2 Answer (see Fig. A.11)

COST OF ON-THE-JOB INJURY CLAIMS

MONTH	COST ($)			
Jan	1,200	Null hypothesis		μ = $16,000
Feb	1,875			
Mar	2,650	Research hypothesis		$\mu \neq$ $16,000
Apr	14,212			
May	11,367	n	12	
Jun	5,126			
Jul	6,875			
Aug	9,482	Mean	9,113	
Sep	10,265			
Oct	19,345			
Nov	24,650	STDEV	7,401	
Dec	2,311			
		s.e.	2,137	
Injury4				
		critical t	2.201	
		t-test	-3.22	

Result: Since the absolute value of -3.22 is greater than the critical t of 2.201, we reject the null hypothesis and accept the research hypothesis

Conclusion: The average cost of on-the-job injury claims this past year was significantly less than the average cost of these claims for the year before last year.

Fig. A.11 Answer to Chap. 4: Practice Problem #2

Chapter 4: Practice Problem #3 Answer (see Fig. A.12)

Chapter 4: Practice problem #3 answer (see Fig. A.12)

DISSOLVED OXYGEN CONTENT(DO) IN MAINE LAKES

DO (mg/L)					
4.6	Null hypothesis:	μ	=	5 mg/L	
4.4					
4.8	Research hypothesis:	μ	≠	5 mg/L	
6.4					
6.5					
6.7	n	16			
6.5					
5.6					
5.4	Mean	5.52			
5.8					
4.9					
5.2	STDEV	0.72			
5.6					
5.7					
5.4	s.e.	0.18			
4.8					
	critical t	2.131			
	t-test	2.87			

Result:	Since the absolute value of 2.87 is greater than the critical t of 2.131, we reject the null hypothesis and accept the research hypothesis.
Conclusion:	The average level of DO in the Maine named lakes is significantly greater than 5 mg/L, and is probably closer to 5.5 mg/L

Fig. A.12 Answer to Chap. 4: Practice Problem #3

Chapter 5: Practice Problem #1 Answer (see Fig. A.13)

Chapter 5: Practice problem #1 answer (see Fig. A.13)

WING LENGTH (mm) OF A SPECIES OF MOSQUITOES IN TWO REGIONS

Group	n	Mean	STDEV
1 North	124	3.2	1.2
2 South	135	3.4	1.3

Null hypothesis:	μ_1	$=$	μ_2
Research hypothesis:	μ_1	$\neq$	μ_2
STDEV1 squared / n1		0.012	
STDEV2 squared / n2		0.013	
E13 + E16		0.024	
s.e.		0.155	
critical t		1.96	
t-test		-1.287	
Result:	Since the absolute value of − 1.287 is less than the critical t value of 1.96, we accept the null hypothesis.		
Conclusion:	There was no difference in wing length of the species of mosquitoes in the North vs. the South regions.		

Fig. A.13 Answer to Chap. 5: Practice Problem #1

Chapter 5: Practice Problem #2 Answer (see Fig. A.14)

Chapter 5: Practice problem #2 answer (see Fig.A.14)

SEDIMENT IN RIVERS IN URBAN AND RURAL SITES

Amount of sediment (mg/L)

urban	rural				
45	35	Null hypothesis:	μ_1	=	μ_2
62	40	Research hypothesis:	μ_1	≠	μ_2
84	55				
95	78	Group	n	Mean	STDEV
55	38	1 Urban	12	76.08	18.74
59	42	2 Rural	12	50.42	13.99
64	48				
94	52				
105	70	(n1 - 1) x STDEV1 squared			3862.92
87	65				
76	44	(n2-1) x STDEV2 squared			2152.92
87	38				
		n1 + n2 - 2			22
		1/n1 + 1/n2			0.17
		s.e.			6.75
		critical t			2.074
		t-test			3.80

Result: Since the absolute value of 3.80 is greater than the critical t of 2.074, we reject the null hypothesis and accept the research hypothesis.

Conclusion: Urban rivers had significantly more sediment in them than Rural rivers (76 mg/L vs. 50 mg/L)

Fig. A.14 Answer to Chap. 5: Practice Problem #2

Chapter 5: Practice Problem #3 Answer (see Fig. A.15)

	WHITE PINE PULPWOOD VOLUME						
	TYPE A	**TYPE B**	Null hypothesis			μ_1 =	μ_2
	6	7					
	8	14	Research hypothesis			$\mu_1 \neq$	μ_2
	9	12					
	7	13	Group	n	Mean	STDEV	
	6	8	1 Type A	16	7.88	1.75	
	10	9	2 Type B	16	10.06	2.74	
	12	10					
	6	8					
	7	7	(n1 - 1) x STDEV1 squared			45.75	
	9	6					
	7	12	(n2 - 1) x STDEV2 squared			112.94	
	6	14					
	7	8	n1 + n2 − 2			30	
	8	14					
	10	9	1/n1 + 1/n2			0.13	
	8	10					
			s.e.			0.81	
			critical t			2.042	
			t-test			-2.69	

Result: Since − 2.69 is greater than the critical t of 2.042, we reject the null hypothesis and accept the research hypothesis

Conclusion: White Pine Type B had significantly more pulpwood volume than White Pine Type A (10.06 vs. 7.88)

Fig. A.15 Answer to Chap. 5: Practice Problem #3

Chapter 6: Practice Problem #1 Answer (see Fig. A.16)

Chapter 6: Practice problem #1 answer (see Fig. A.16)

Research question: "What is the relationship between the weight of a 4-door sedan and its miles per gallon (mpg) performance in city driving?"

Weight (1000 lbs)	City Miles Per Gallon (mpg)
2.1	32.2
2.4	28.6
3.5	26.7
2.3	28.1
3.4	27.7
4.1	16.2
3.8	20.9
3.6	22.4
4.3	18.4
4.2	15.3

RELATIONSHIP BETWEEN WEIGHT AND CITY mpg IN 4-DOOR SEDANS

$y = -6.3938x + 45.197$

SUMMARY OUTPUT

Regression Statistics	
Multiple R	0.900
R Square	0.8
Adjusted R Square	0.8
Standard Error	2.7
Observations	10

ANOVA

	df	SS	MS	F	Significance F
Regression	1	247.0	247.0	34.0	0.0004
Residual	8	58.1	7.3		
Total	9	305.0			

	Coefficients	Standard Error	t Stat	P-value	Lower 95%	Upper 95%
Intercept	45.197	3.8	11.9	0.0	36.5	53.9
X Variable 1	-6.394	1.1	-5.8	0.0	-8.9	-3.9

Fig. A.16 Answer to Chap. 6: Practice Problem #1

Chapter 6: Practice Problem #1 (continued)

(d) a = y-intercept = +45.197
       b = slope = −6.394
(e) Y = a + b X
       Y = 45.197 − 6.394 X
(f)   r = −.900
(g) Y = 45.197 − 6.394 (2.5)
       Y = 45.197 − 15.985
       Y = 29.212 mpg
(h) About 22 mpg

Chapter 6: Practice Problem #2 Answer (see Fig. A.17)

Chapter 6: Practice problem #2 answer (see Fig. A.17)

DOWN-HOLE DEPTH (meters) VS. TEMPERATURE (degrees centigrade)

DEPTH (m)	TEMPERATURE ($^{\circ}$C)
0.1	-3.6
0.3	-3.5
0.6	-2.7
0.9	-2.5
1.4	-2.6
2.2	-2.7
3.2	-2.4
4.8	-0.2
6.8	0.0

SUMMARY OUTPUT

Regression Statistics	
Multiple R	0.94
R Square	0.8780
Adjusted R Square	0.8605
Standard Error	0.4808
Observations	9

ANOVA

	df	SS	MS	F	Significance F
Regression	1	11.6439	11.6439	50.3659	0.0002
Residual	7	1.6183	0.2312		
Total	8	13.2622			

	Coefficients	Standard Error	t Stat	P-value	Lower 95%	Upper 95%	Lower 95.0%	Upper 95.0%
Intercept	-3.43	0.2320	-14.8055	0.0000	-3.9835	-2.8863	-3.9835	-2.8863
X Variable 1	0.53	0.0744	7.0969	0.0002	0.3519	0.7036	0.3519	0.7036

Fig. A.17 Answer to Chap. 6: Practice Problem #2

Chapter 6: Practice Problem #2 (continued)
(2b) about −1.9 degrees centigrade

1. $r = +.94$
2. $a = $ y-intercept $= -3.43$
3. $b = $ slope $= +0.53$
4. $Y = a + b\,X$
   $Y = -3.43 + 0.53\,X$
5. $Y = -3.43 + 0.53\,(2)$
   $Y = -3.43 + 1.06$
   $Y = -2.37$ degrees centigrade

Chapter 6: Practice Problem #3 Answer (see Fig. A.18)

RELATIONSHIP BETWEEN AIR POLLUTION LEVEL AND PEAK FLOW IN ADULT WOMEN (ages 30-40)

AIR POLLUTION (μ g/m^3)	PEAK FLOW (ml/sec)
21	52
63	43
45	51
103	46
25	48
34	44
39	42
45	41
43	45
48	47
55	51
89	41
96	35
91	32
102	40

SUMMARY OUTPUT

Regression Statistics	
Multiple R	0.62
R Square	0.388
Adjusted R Square	0.341
Standard Error	4.630
Observations	15

ANOVA

	df	SS	MS	F	Significance F
Regression	1	176.997	176.997	8.255	0.013
Residual	13	278.737	21.441		
Total	14	455.733			

	Coefficients	Standard Error	t Stat	P-value	Lower 95%
Intercept	51.30	2.852	17.991	0.000	45.144
X Variable 1	-0.12	0.043	-2.873	0.013	-0.217

Fig. A.18 Answer to Chap. 6: Practice Problem #3

Chapter 6: Practice Problem #3 (continued)

(d) a = y-intercept = 51.30

    b = slope = −0.12

(e) Y = a + b X

    Y = 51.30 − 0.12 X

(f) r = −.62

(g) Y = 51.30 − 0.12 X

    Y = 51.30 − 0.12 (80)

    Y = 51.30 − 9.6

    Y = 41.7 ml/s

(h) About 45–47 ml/s

Chapter 7: Practice Problem #1 Answer (see Fig. A.19)

Chapter 7: Practice problem #1 answer (see Fig. A.19)

SEED PRODUCTION

How well do water, light, fertilizer, and space predict the number of seeds produced per pod?

AVERAGE SEEDS PER POD	WATER (ml)	LIGHT RECEIVED (min)	SPACE (cm)	FERTILIZER (µl)
3.25	600	620	12.5	650
3.42	520	550	10	600
2.85	510	540	5	500
2.65	480	460	2.5	510
3.65	720	710	15	630
3.16	570	610	7.5	550
3.56	710	650	10	610
2.35	500	480	5	430
2.86	450	470	7.5	450
2.95	560	530	10	550
3.15	550	580	10	580
3.45	610	620	12.5	620

SUMMARY OUTPUT

Regression Statistics	
Multiple R	0.93
R Square	0.857875018
Adjusted R Square	0.776660743
Standard Error	0.184379009
Observations	12

ANOVA

	df	SS	MS	F	Significance F
Regression	4	1.436397334	0.359099334	10.56310622	0.004331962
Residual	7	0.237969332	0.033995619		
Total	11	1.674366667			

	Coefficients	Standard Error	t Stat	P-value	Lower 95%
Intercept	0.5682	0.632835292	0.897786192	0.399124683	-0.928268892
WATER (ml)	-0.0004	0.00181073	-0.210590788	0.83920554	-0.00466302
LIGHT RECEIVED (min)	0.0022	0.002425544	0.907419349	0.39434945	-0.003528685
SPACE (cm)	0.0200	0.029379257	0.682445079	0.516897071	-0.049421175
FERTILIZER (µl)	0.0024	0.001545116	1.543035862	0.166730022	-0.001269449

	AVERAGE SEEDS PER POD	WATER (ml)	LIGHT RECEIVED (min)	SPACE (cm)	FERTILIZER (µl)
AVERAGE SEEDS PER POD	1				
WATER (ml)	0.79	1			
LIGHT RECEIVED (min)	0.87	0.93	1		
SPACE (cm)	0.83	0.74	0.82	1	
FERTILIZER (µl)	0.89	0.76	0.83	0.81	1

Fig. A.19 Answer to Chap. 7: Practice Problem #1

Chapter 7: Practice Problem #1 (continued)

1. Multiple correlation $= .93$
2. y-intercept $= 0.5682$
3. $b_1 = -0.0004$
4. $b_2 = 0.0022$
5. $b_3 = 0.0200$
6. $b_4 = 0.0024$
7. $Y = a + b_1 X_1 + b_2 X_2 + b_3 X_3 + b_4 X_4$
   $Y = 0.5682 - 0.0004 X_1 + 0.0022 X_2 + 0.0200 X_3 + 0.0024 X_4$
8. $Y = 0.5682 - 0.0004 (610) + 0.0022 (550) + 0.0200 (3) + 0.0024 (610)$
   $Y = 0.5682 - 0.244 + 1.21 + 0.06 + 1.464$
   $Y = 3.302 - 0.244$
   $Y = 3.06$ seeds per pod
9. 0.79
10. 0.87
11. 0.83
12. 0.89
13. 0.74
14. 0.83
15. The best predictor of AVERAGE SEEDS PER POD was FERTILIZER $(r = .89)$.
16. The four predictors combined predict AVERAGE SEEDS PER POD at $R_{xy} = .93$, and this is much better than the best single predictor by itself.

Chapter 7: Practice Problem #2 Answer (see Fig. A.20)

GRADUATE RECORD EXAMINATIONS (GRE)

FIRST-YEAR GPA	GRE QUANTITATIVE	GRE WRITING
3.25	160	5
3.42	156	4
2.85	156	2
2.65	154	1
3.65	166	6
3.16	159	3
3.56	166	4
2.35	155	2
2.86	153	3
2.95	158	4
3.15	158	4
3.45	160	5
3.54	165	6
3.36	163	5
3.15	159	4

SUMMARY OUTPUT

Regression Statistics	
Multiple R	0.88
R Square	0.781
Adjusted R Square	0.745
Standard Error	0.186
Observations	15

ANOVA

	df	SS	MS	F	Significance F
Regression	2	1.477	0.738	21.410	0.000
Residual	12	0.414	0.034		
Total	14	1.891			

	Coefficients	Standard Error	t Stat	P-value	Lower 95%
Intercept	-2.69	2.861	-0.941	0.365	-8.925
GRE QUANTITATIVE	0.03	0.019	1.757	0.104	-0.008
GRE WRITING	0.14	0.055	2.533	0.026	0.019

	FIRST-YEAR GPA	GRE QUANTITATIVE	GRE WRITING
FIRST-YEAR GPA	1.00		
GRE QUANTITATIVE	0.81	1.00	
GRE WRITING	0.85	0.78	1.00

Fig. A.20 Answer to Chap. 7: Practice Problem #2

Chapter 7: Practice Problem #2 (continued)

Let $X_1 = $ GRE QUANTITATIVE and $X_2 = $ GRE WRITING

(e) (1a) $R_{xy} = .88$
    (2b) $a = $ y-intercept $= -2.69$
    $b_1 = 0.03$
    $b_2 = 0.14$
(f) $Y = a + b_1 X_1 + b_2 X_2$
    $Y = -2.69 + 0.03 X_1 + 0.14 X_2$
(g) $Y = -2.69 + 0.03 (157) + 0.14 (4)$
    $Y = -2.69 + 4.71 + 0.56$
    $Y = 2.58$ GPA

(k) (1a) $+.81$
    (2b) $+.85$
    (3c) $+.78$
    (4d) The better predictor of GPA is GRE WRITING $(r = +.85)$
    (5e) The two predictors combined predict GPA much better $(R_{xy} = .88)$ than the better single predictor by itself

Chapter 7: Practice Problem #3 Answer (see Fig. A.21)

Chapter 7: Practice problem #3 answer (see Fig. A.21)

CORN YIELDS ON IOWA FARMS

YIELD (bu/acre)	RAINFALL (in/yr)	NITROGEN (lbs/acre)	TEMPERATURE (degrees F)
240	28.8	190	80
255	29.5	195	79
299	27.6	230	79
230	28.3	175	78
300	31.5	250	79
180	27.1	250	82
294	28.9	220	78
278	27.1	200	74
284	28.8	204	80
194	30.7	185	81
170	27.9	190	88

SUMMARY OUTPUT

Regression Statistics	
Multiple R	0.78
R Square	0.602
Adjusted R Square	0.431
Standard Error	36.701
Observations	11

ANOVA

	df	SS	MS	F
Regression	3	14247.810	4749.270	3.526
Residual	7	9428.735	1346.962	
Total	10	23676.545		

	Coefficients	Standard Error	t Stat	P-value
Intercept	744.95	375.630	1.983	0.088
RAINFALL (in/yr)	6.33	8.283	0.764	0.470
NITROGEN (lbs/acre)	0.51	0.450	1.137	0.293
TEMPERATURE (degrees F)	-9.84	3.417	-2.880	0.024

	YIELD (bu/acre)	RAINFALL (in/yr)	NITROGEN (lbs/acre)	TEMPERATURE (degrees F)
YIELD (bu/acre)	1			
RAINFALL (in/yr)	0.19	1		
NITROGEN (lbs/acre)	0.31	0.03	1	
TEMPERATURE (degrees F)	-0.70	0.00	-0.05	1

Fig. A.21 Answer to Chap. 7: Practice Problem #3

Chapter 7: Practice Problem #3 (continued)

1. Multiple correlation $= .78$
2. $a = $ y-intercept $= 744.95$
3. $b_1 = 6.33$
4. $b_2 = 0.51$
5. $b_3 = -9.84$
6. $Y = a + b_1 X_1 + b_2 X_2 + b_3 X_3$
   $Y = 744.95 + 6.33 X_1 + 0.51 X_2 - 9.84 X_3$
7. $Y = 744.95 + 6.33 (28) + 0.51 (205) - 9.84 (83)$
   $Y = 744.95 + 177.24 + 104.55 - 816.72$
   $Y = 1026.74 - 816.72 = 210$ bu/acre
8. $+0.19$
9. $+0.31$
10. -0.70
11. $+0.03$
12. $+0.00$
13. -0.05
14. The best single predictor of corn yield was TEMPERATURE ($r = -.70$). *(Note: Remember to ignore the negative sign and just use 0.70).*
15. The three predictors combined predict corn yield much better at $R_{xy} = .78$, and this is much better than the best single predictor by itself.

Chapter 8: Practice Problem #1 Answer (see Fig. A.22)

Chapter 8: Practice problem #1 answer (see Fig. A.22)

CHEMICAL REACTION YIELD (grams of product produced)

BELOW ROOM TEMP (15 °C)	ROOM TEMP (25 °C)	ABOVE ROOM TEMP (30 °C)
90	85	76
85	89	80
74	83	90
89	79	84
84	74	78
95	75	65
92	86	42
65	87	58
75	86	63
73	88	75
54		66
71		

Anova: Single Factor

SUMMARY

Groups	Count	Sum	Average	Variance
BELOW ROOM TEMP (15 °C)	12	947	78.92	151.72
ROOM TEMP (25 °C)	10	832	83.20	28.84
ABOVE ROOM TEMP (30 °C)	11	777	70.64	183.45

ANOVA

Source of Variation	SS	df	MS	F	P-value	F crit
Between Groups	867.12	2	433.56	3.46	0.04	3.32
Within Groups	3763.06	30	125.44			
Total	4630.18	32				

ROOM TEMP vs. ABOVE ROOM TEMP

1/n ROOM TEMP + 1/n ABOVE ROOM TEMP	0.19
s. e. ROOM TEMP vs. ABOVE ROOM TEMP	4.89
ANOVA t-test	2.57

Fig. A.22 Answer to Chap. 8: Practice Problem #1

Chapter 8: Practice Problem #1 (continued)

Let Group 1 = BELOW ROOM TEMP, Group 2 = ROOM TEMP, and Group 3 = ABOVE ROOM TEMP

1. H_0: $\mu_1 = \mu_2 = \mu_3$
   H_1: $\mu_1 \neq \mu_2 \neq \mu_3$
2. $MS_b = 433.56$
3. $MS_w = 125.44$
4. $F = 433.56/125.44 = 3.46$
5. critical $F = 3.32$
6. Result: Since 3.46 is greater than 3.32, we reject the null hypothesis and accept the research hypothesis.
7. There was a significant difference between the three temperatures in the grams of product produced.
   ROOM TEMP vs. ABOVE ROOM TEMP
8. H_0: $\mu_2 = \mu_3$
   H_1: $\mu_2 \neq \mu_3$
9. 83.20
10. 70.64
11. $df = 33 - 3 = 30$
12. critical $t = 2.042$
13. $1/10 + 1/11 = 0.10 + 0.09 = 0.19$
    s.e. = SQRT ($125.44 * 0.19$) = SQRT (23.83) = 4.88
14. ANOVA $t = (83.20 - 70.64)/4.88 = 2.57$
15. Result: Since the absolute value of 2.57 is greater than 2.042, we reject the null hypothesis and accept the research hypothesis.
16. Conclusion: ROOM TEMP produced significantly more grams of product than ABOVE ROOM TEMP (83.2 vs. 70.6).

Chapter 8: Practice Problem #2 Answer (see Fig. A.23)

Chapter 8: Practice problem #2 answer (see Fig. A.23)

DAISY FLEABANE GROWTH IN ELEVATED NITROGEN SOILS

	No Nitrogen	Low Nitrogen	High Nitrogen
	500	550	800
	550	600	1000
	700	750	900
	650	700	1100
	500	600	1400
	550	650	1600
	600	600	1500
	650	700	1800
	600	650	1500
		750	1800
			2000
			1600

Anova: Single Factor

SUMMARY

Groups	Count	Sum	Average	Variance
No Nitrogen	9	5300	588.89	4861.11
Low Nitrogen	10	6550	655.00	4694.44
High Nitrogen	12	17000	1416.67	148787.88

ANOVA

Source of Variation	SS	df	MS	F	P-value	F crit
Between Groups	4645581.54	2	2322790.77	37.86	1.09E-08	3.34
Within Groups	1717805.56	28	61350.20			
Total	6363387.10	30				

Low Nitrogen vs. High Nitrogen

1/10 + 1/12	0.18
s.e. ANOVA	106.05
ANOVA t-test	-7.18

Fig. A.23 Answer to Chap. 8: Practice Problem #2

Chapter 8: Practice Problem #2 (continued)

Let Group 1 = No Nitrogen, Group 2 = Low Nitrogen, and Group 3 = High Nitrogen

1. Null hypothesis: $\mu_1 = \mu_2 = \mu_3$
   Research hypothesis: $\mu_1 \neq \mu_2 \neq \mu_3$
2. $MS_b = 2{,}322{,}790.77$
3. $MS_w = 61{,}350.20$
4. $F = 2{,}322{,}790/61{,}350 = 37.86$
5. critical $F = 3.34$
6. Since the F-value of 37.86 is greater than the critical F value of 3.34, we reject the null hypothesis and accept the research hypothesis.
7. There was a significant difference between the weight of Daisy Fleabane between the three treatments.
   Treatment 2 vs. Treatment 3
8. $H_0: \mu_2 = \mu_3$
   $H_1: \mu_2 \neq \mu_3$
9. 655
10. 1416.67
11. $df = 31 - 3 = 28$
12. critical $t = 2.048$
13. $1/10 + 1/12 = 0.10 + 0.08 = 0.18$
    s.e. $= SQRT\ (61{,}350.20*0.18) = SQRT\ (11{,}043.04) = 105.086$
14. ANOVA $t = (655 - 1416.67)/105.086 = -761.67/105.086 = -7.25$
15. Result: Since the absolute value of -7.25 is greater than 2.048, we reject the null hypothesis and accept the research hypothesis.
16. Conclusion: Daisy Fleabane flowers weighed significantly more when the soil contained High Nitrogen than when the soil contained Low Nitrogen (1417 mg vs. 655 mg).

Chapter 8: Practice Problem #3 Answer (see Fig. A.24)

DISSOLVED OXYGEN CONTENT IN WATER (ppm)

	LOCATION	
A	B	C
6.2	3.8	3.9
6.9	4.9	4.9
6.8	3.9	4.1
6.6	4.1	4.3
6.5	4.6	4.6
6.4	4.2	4.5
6.9	4.6	4.7
6.7	4.7	4.8
6.5	4.8	4.7
6.8	4.4	4.6

Anova: Single Factor

SUMMARY

Groups	Count	Sum	Average	Variance
A	10	66.30	6.63	0.05
B	10	44.00	4.40	0.15
C	10	45.10	4.51	0.10

ANOVA

Source of Variation	SS	df	MS	F	P-value	F crit
Between Groups	31.60	2	15.80	157.41	0.00	3.35
Within Groups	2.71	27	0.10			
Total	34.31	29				

LOCATION A vs. B

1/n A + 1/n B	0.20
s.e.	0.14
ANOVA t	15.74

Fig. A.24   Answer to Chap. 8: Practice Problem #3

Chapter 8: Practice Problem #3 (continued)

1. Null hypothesis: $\mu_A = \mu_B = \mu_C$
   Research hypothesis: $\mu_A \neq \mu_B \neq \mu_C$
2. $MS_b = 15.80$
3. $MS_w = 0.10$
4. $F = 157.41$
5. critical $F = 3.35$
6. Result: Since the F-value of 157.41 is greater than the critical F value of 3.35, we reject the null hypothesis and accept the research hypothesis.
7. Conclusion: There was a significant difference between the three locations in dissolved oxygen content.
8. Null hypothesis: $\mu_A = \mu_B$
   Research hypothesis: $\mu_A \neq \mu_B$
9. 6.63
10. 4.40
11. degrees of freedom $= 30 - 3 = 27$
12. critical $t = 2.052$
13. s.e. $_{ANOVA} = 0.14$
14. ANOVA $t = 15.74$
15. Since the absolute value of 15.74 is greater than the critical t of 2.052, we reject the null hypothesis and accept the research hypothesis.
16. LOCATION A had significantly more dissolved oxygen content than LOCATION B (6.63 vs. 4.40).

Appendix B: Practice Test

Chapter 1: Practice Test

Air pollution is a serious health problem in major cities of the world. As the Ozone level of the air increases, it makes it more difficult for people to breathe normally, and this can have long-term effects on their health and welfare. A geographical area with "ambient" levels of Ozone has a low level of air pollution. Suppose that you measured the Ozone level in various parts of San Francisco, and that you wanted to determine the city's average Ozone concentration in its air. The hypothetical data measured in parts per billion (ppb) are given in Fig. B.1:

Fig. B.1 Worksheet Data
for Chap. 1 Practice Test
(Practical Example)

AMBIENT AIR OZONE CONTRATIONS IN SAN FRANCISCO

parts per billion (ppb)
4
6
5
7
9
8
10
5
7
6
4
9
4
10
8
7
5
6

(a) Create an Excel table for these data, and then use Excel to the right of the table to find the sample size, mean, standard deviation, and standard error of the mean for these data. Label your answers and round off the mean, standard deviation, and standard error of the mean to two decimal places.

(b) Save the file as: OZONE3

Chapter 2: Practice Test

Suppose that you have been asked to select a random sample of 22 plots in a farm field to test the soil for its nutrient levels. The field has been mapped out carefully into 215 separate plots, all of which have the same area in square feet.

(a) Set up a spreadsheet of frame numbers for these plots with the heading: FRAME NUMBERS
(b) Then, create a separate column to the right of these frame numbers which duplicates these frame numbers with the title: Duplicate frame numbers.
(c) Then, create a separate column to the right of these duplicate frame numbers called RAND NO. and use the =RAND() function to assign random numbers to all of the frame numbers in the duplicate frame numbers column and change this column format so that three decimal places appear for each random number.
(d) Sort the *duplicate frame numbers and random numbers* into a random order.
(e) Print the result so that the spreadsheet fits onto one page.
(f) Circle on your printout the I.D. number of the first 22 plots that you would use in your soil test.

Save the file as: RAND18

Important note: Note that everyone who does this problem will generate a different random order of plots ID numbers since Excel assigns a different random number each time the RAND() command is used. For this reason, the answer to this problem given in this Excel Guide will have a completely different sequence of random numbers from the random sequence that you generate. This is normal and what is to be expected.

Chapter 3: Practice Test

Suppose that an environmental scientist wants to determine if the percent of sand content in the soil of a particular field averages 57%. Suppose, further, that this scientist has given you the sand content data from a random sample of grids of the same length and width in this field (1 × 1-meter, m). You have been asked to "run the data" to see if the average sand content of this field is 57%, and you have decided to test your Excel skills on the hypothetical data given in Fig. B.2

SAND CONTENT (%) FROM SOIL SAMPLES TAKEN FROM DIFFERENT LOCATIONS IN A FIELD

SAND CONTENT (%)
52.4
55.6
58.1
60.2
62.4
63.7
51.9
52.7
62.5
58.4
56.9
60.4
61.5

Fig. B.2 Worksheet Data for Chap. 3 Practice Test (Practical Example)

(a) Create an Excel table for these data, and use Excel to the right of the table to find the sample size, mean, standard deviation, and standard error of the mean for these data. Label your answers and round off the mean, standard deviation, and standard error of the mean to two decimal places in number format.

(b) By hand, write the null hypothesis and the research hypothesis on your printout.

(c) Use Excel's *TINV function* to find the 95% confidence interval about the mean for these data. Label your answers. Use two decimal places for the confidence interval figures in number format.

(d) On your printout, draw a diagram of this 95% confidence interval by hand, including the reference value.

(e) On your spreadsheet, enter the *result.*

(f) On your spreadsheet, enter the *conclusion in plain English.*

(g) Print the data and the results so that your spreadsheet fits onto one page.

(h) Save the file as: SAND3

Chapter 4: Practice Test

Suppose that you work for a company that manufactures small submersible pumps. Submersible pumps are pumps that can be submerged under water and they are used to pump water out of an area. For example, submersible pumps can be used to pump flood water out of basements. Suppose, further, that your company has developed a new style of pump and has decided to test it on some recently flooded homes near Grafton, Illinois, in the USA. The old style pumps pumped an average of 46 gallons per minute (gal/min). You want to test your Excel skills on a small sample of data using your company's new submersible pumps using the hypothetical data given in Fig. B.3.

Fig. B.3 Worksheet Data
for Chap. 4 Practice Test
(Practical Example)

OUTPUT OF NEW SUBMERSIBLE PUMP

GALLONS PER MINUTE (gal/min)
51
50
50
49
50
48
52
50
50
49
48
49
50
51
49
50
49
51
51
50

(a) Write the null hypothesis and the research hypothesis on your spreadsheet.
(b) Create a spreadsheet for these data, and then use Excel to find the sample size, mean, standard deviation, and standard error of the mean to the right of the data set. Use number format (two decimal places) for the mean, standard deviation, and standard error of the mean.
(c) Type the *critical t* from the t-table in Appendix E onto your spreadsheet and label it.
(d) Use Excel to compute the t-test value for these data (use two decimal places) and label it on your spreadsheet.
(e) Type the *result* on your spreadsheet and then type the *conclusion in plain English* on your spreadsheet.
(f) Save the file as: PUMP8

Chapter 5: Practice Test

Suppose that you wanted to study the duration of hibernation of a species of hedgehogs (*Erinaceus europaeus*) in two regions of the USA (NORTH vs. SOUTH). Suppose, further, that researchers have captured hedgehogs in these regions, attached radio tags to their bodies, and then released them back into the site where they were captured. The researchers monitored the movements of the hedge-hogs during the winter months to determine the number of days that they did not leave their nests. The researchers have selected a random sample of hedgehogs from each region, and you want to test your Excel skills on the hypothetical data given in Fig. B.4.

Fig. B.4 Worksheet Data
for Chap. 5 Practice Test
(Practical Example)

NUMBER OF DAYS SPENT IN THE NEST

NORTH REGION	SOUTH REGION
95	90
98	92
99	94
112	93
103	95
105	98
106	100
109	102
110	103
111	104
112	98
114	102
95	99
98	
103	

(a) Write the null hypothesis and the research hypothesis.
(b) Create an Excel table that summarizes these data.
(c) Use Excel to find the standard error of the difference of the means.
(d) Use Excel to perform a *two-group t-test*. What is the value of *t* that you obtain (use two decimal places)?
(e) On your spreadsheet, type the *critical value of t* using the t-table in Appendix E.
(f) Type the *result* of the test on your spreadsheet.
(g) Type your *conclusion in plain English* on your spreadsheet.
(h) Save the file as: HEDGE3
(i) Print the final spreadsheet so that it fits onto one page.

Chapter 6: Practice Test

Suppose that you worked as a manager in a large county government office and that you wanted to encourage small cities and municipalities within your county to create new recycling centers in their localities to make it easier for people who lived there to recycle their paper and plastic waste by not having to drive a long distance to the nearest recycling center in the county. Your hypothesis is that the farther the recycling center is from the homes of the residents, the less likely they will be to use these centers in a typical year. You have decided to interview people at the various recycling centers in the county that already exist, and to ask them how far they lived from that recycling center so that you could check the distance on Google maps. Before conducting these interviews, you want to check to make sure that you can do the data analysis correctly resulting from these interviews, and so you have created the hypothetical data that appear in Fig. B.5 below:

Research question:	"Does the distance from the nearest recycling center affect the frequency of visiting that recycling center?

DISTANCE (km)	FREQUENCY (days/year)
1	55
2	25
2.5	35
4	30
5	20
6	25
4	28
7	15
7.5	12
5.5	17
4.5	26
1.5	40
2.5	30
3.5	35

Fig. B.5 Worksheet Data for Chap. 6 Practice Test (Practical Example)

Create an Excel spreadsheet and enter the data.

(a) Create an *XY scatterplot* of these two sets of data such that:

- top title: RELATIONSHIP BETWEEN DISTANCE AND FREQUENCY OF RECYCLING
- x-axis title: DISTANCE (km)
- y-axis title: FREQUENCY (days/year)

- move the chart below the table
- re-size the chart so that it is 7 columns wide and 25 rows long
- delete the legend
- delete the gridlines

(b) Create the *least-squares regression line* for these data on the scatterplot.
(c) Use Excel to run the regression statistics to find the *equation for the least-squares regression line* for these data and display the results below the chart on your spreadsheet. Add the regression equation to the chart. Use number format (two decimal places) for the correlation and for the coefficients.

Print *just the input data and the chart* so that this information fits onto one page in portrait format.

Then, print *just the regression output table* on a separate page so that it fits onto that separate page in portrait format.

By hand:

(d) Circle and label the value of the *y-intercept* and the *slope* of the regression line on your printout.
(e) Write the regression equation *by hand* on your printout for these data (use two decimal places for the y-intercept and the slope).
(f) Circle and label the *correlation* between the two sets of scores in the regression analysis summary output table on your printout.
(g) Underneath the regression equation you wrote by hand on your printout, use the regression equation to predict the average frequency of visits to a recycling center for residents who lived 4 kilometers (km) from that center.
(h) *Read from the graph,* the average frequency you would predict for a resident who lived 6 kilometers (km) from the recycling center, and write your answer in the space immediately below:

_____

(i) Save the file as: DISTANCE3

Chapter 7: Practice Test

Suppose that you wanted to estimate the total number of gallons required for the latest model 4-door sedans when they were driven on a specific route of 200 miles between St. Louis, Missouri, and Indianapolis, Indiana, at specified speeds using drivers that were about the same weight. You have decided to use two predictors: (1) weight of the car (measured in thousands of pounds), and (2) the car's engine horsepower. To check your skills in Excel, you have created the hypothetical data given in Fig. B.6.

TOTAL GALLONS USED TO DRIVE FROM ST. LOUIS TO INDIANAPOLIS

FOUR-DOOR SEDANS

TOTAL GALLONS USED	WEIGHT (1000 lbs)	HORSEPOWER
6.1	3.8	130
6.3	3.7	150
4.8	4.0	140
4.2	2.4	125
3.8	2.9	98
4.7	3.0	115
3.5	2.1	121
5.5	2.9	123
5.9	3.1	110
3.4	2.1	96

Fig. B.6 Worksheet Data for Chap. 7 Practice Test (Practical Example)

(a) Create an Excel spreadsheet using TOTAL GALLONS USED as the criterion (Y), and the other variables as the two predictors of this criterion (X_1 = WEIGHT (1000 lbs), and X_2 = HORSEPOWER).

(b) Use Excel's *multiple regression* function to find the relationship between these three variables and place the SUMMARY OUTPUT below the table.

(c) Use number format (two decimal places) for the multiple correlation on the Summary Output, and use two decimal places for the coefficients in the SUMMARY OUTPUT.

(d) Save the file as: GALLONS9

(e) Print the table and regression results below the table so that they fit onto one page.

Answer the following questions using your Excel printout:

1. What is the multiple correlation R_{xy}?
2. What is the y-intercept a?
3. What is the coefficient for WEIGHT b_1?
4. What is the coefficient for HORSEPOWER b_2?
5. What is the multiple regression equation?
6. Predict the TOTAL GALLONS USED you would expect for a WEIGHT of 3,800 pounds and a car that had 126 HORSEPOWER.

(f) Now, go back to your Excel file and create a correlation matrix for these three variables and place it underneath the SUMMARY OUTPUT.

(g) Re-save this file as: GALLONS9

(h) Now, print out *just this correlation matrix* on a separate sheet of paper.

Answer to the following questions using your Excel printout. (Be sure to include the plus or minus sign for each correlation):

7. What is the correlation between WEIGHT and TOTAL GALLONS USED?
8. What is the correlation between HORSEPOWER and TOTAL GALLONS USED?
9. What is the correlation between WEIGHT and HORSEPOWER?
10. Discuss which of the two predictors is the better predictor of total gallons used.
11. Explain in words how much better the two predictor variables combined predict total gallons used than the better single predictor by itself.

Chapter 8: Practice Test

Nitrogen dioxide (NO_2) is a reddish-brown toxic gas that has a sharp odor and is a prominent source of air pollution in the exhaust fumes of motor vehicles that have internal combustion engines. It is typically created by the oxidation of nitric oxide by oxygen in the air.

Suppose that London, England, has been running a pilot test of two types of vehicles that could be used in its public transit systems: (1) a special type of minibus that can carry nine to sixteen seated passengers, and (2) a large-scale bus that has been adapted to have a special engine that emits less NO_2 into the atmosphere than its current double-decker red buses, and that you have been hired to compare the NO_2 emissions of passenger cars, these types of minibuses, and these types of buses. You have collected data from a sample of these types of vehicles in which NO_2 was measured in parts per billion (ppb), and you want to test your Excel skills on the hypothetical data given in Fig. B.7:

NO$_2$ CONCENTRATION (ppb) IN THE EXHAUST FUMES OF VEHICLES

CAR	MINIBUS	BUS
63	46	38
65	48	40
71	47	41
72	49	43
69	50	44
63	52	46
65	53	47
66	56	42
67	58	48
71	54	37
70	57	36
68		38
		39
		40

Fig. B.7 Worksheet Data for Chap. 8 Practice Test (Practical Example)

(a) Enter these data on an Excel spreadsheet.
Let CAR = Group 1, MINIBUS = Group 2, and BUS = Group 3.
(b) On your spreadsheet, write the null hypothesis and the research hypothesis for these data.
(c) Perform a *one-way ANOVA test* on these data and show the resulting ANOVA table underneath the input data for the three types of vehicles.
(d) If the F-value in the ANOVA table is significant, create an Excel formula to compute the ANOVA t-test comparing CARS versus BUSES, and show the results below the ANOVA table on the spreadsheet (put the standard error and the ANOVA t-test value on separate lines of your spreadsheet, and use two decimal places for each value).
(e) Print out the resulting spreadsheet so that all of the information fits onto one page.
(f) On your printout, label by hand the MS (between groups) and the MS (within groups).
(g) Circle and label the value for F on your printout for the ANOVA of the input data.
(h) Label by hand on the printout the mean for CARS and the mean for BUSES that were produced by your ANOVA formulas.

(i) Save the spreadsheet as: FUMES3

On a separate sheet of paper, now do the following by hand:

(j) Find the critical value of F in the ANOVA Single Factor results table.

(k) Write a summary of the *result* of the ANOVA test for the input data.

(l) Write a summary of the *conclusion* of the ANOVA test in plain English for the input data.

(m) Write the null hypothesis and the research hypothesis comparing CARS versus BUSES.

(n) Compute the degrees of freedom for the *ANOVA t-test* by hand for three types of vehicles.

(o) Use your calculator and Excel to compute the standard error (s.e.) of the ANOVA t-test.

(p) Use your calculator and Excel to compute the ANOVA t-test value.

(q) Write the *critical value of t* for the ANOVA t-test using the table in Appendix E.

(r) Write a summary of the *result* of the ANOVA t-test.

(s) Write a summary of the *conclusion* of the ANOVA t-test in plain English.

Appendix C: Answers to Practice Test

Practice Test Answer: Chapter 1 (see. Fig. C.1)

AMBIENT AIR OZONE CONTRATIONS IN SAN FRANCISCO		
parts per billion (ppb)		
4		
6		
5	n	18
7		
9		
8	MEAN	6.67
10		
5		
7	STDEV	2.00
6		
4		
9	s.e.	0.47
4		
10		
8		
7		
5		
6		

Fig. C.1 Practice Test Answer to Chap. 1 Problem

Practice Test Answer: Chapter 2 (see. Fig. C.2)

FRAME NUMBERS	Duplicate frame numbers	RAND NO.
1	26	0.195
2	174	0.599
3	2	0.200
4	74	0.776
5	207	0.138
6	173	0.274
7	105	0.661
8	30	0.045
9	170	0.753
10	41	0.808
11	158	0.867
12	9	0.033
13	154	0.625
14	100	0.214
15	138	0.157
16	24	0.514
17	140	0.031
18	53	0.594
19	151	0.581
20	83	0.340
21	150	0.513
22	19	0.376
23	34	0.956
24	122	0.917
25	89	0.178
26	181	0.968
27	31	0.788
28	96	0.426
29	65	0.607
30	15	0.183
31		309
	12	
209	145	0.00/
210	37	0.198
211	72	0.610
212	52	0.312
213	129	0.780
214	126	0.234
215	6	0.201

Fig. C.2 Practice Test Answer to Chap. 2 Problem

Practice Test Answer: Chapter 3 (see. Fig. C.3)

			Null hypothesis:	μ	=	57 percent
			Research hypothesis:	μ	≠	57 percent

SAND CONTENT (%) FROM SOIL SAMPLES TAKEN FROM DIFFERENT LOCATIONS IN A FIELD

SAND CONTENT (%)			
52.4	n	13	
55.6			
58.1			
60.2	Mean	58.21	
62.4			
63.7			
51.9	STDEV	4.06	
52.7			
62.5			
58.4	s.e.	1.13	
56.9			
60.4			
61.5	95% confidence interval		

	Lower limit:	55.75
	Upper limit:	60.66

Draw a picture below the confidence interval

55.75--------- ------57---- ------58.21-------- ---60.66
lower   Ref.   mean   upper
limit   Value   limit

Result:	Since the reference value of 57% is inside the confidence interval, we accept the null hypothesis.
Conclusion:	The average sand content of this field is 57 percent.

Fig. C.3 Practice Test Answer to Chap. 3 Problem

Practice Test Answer: Chapter 4 (see. Fig. C.4)

OUTPUT OF NEW SUBMERSIBLE PUMP					
	GALLONS PER MINUTE (gal/min)				
	51				
	50				
	50	Null hypothesis:	μ	=	46 gal/min
	49				
	50	Research hypothesis:	μ	≠	46 gal/min
	48				
	52				
	50	n	20		
	50				
	49				
	48	Mean	49.85		
	49				
	50				
	51	STDEV	1.04		
	49				
	50				
	49	s.e.	0.23		
	51				
	51				
	50	critical t	2.093		
		t-test	16.56		

Result: Since the absolute value of 16.56 is greater than the critical t of 2.093,
we reject the null hypothesis and accept the research hypothesis.

Conclusion: The new style of submersible pumps pumped significantly more water
than the older style of pumps, and it was probably closer to almost 50 gal/min.

Fig. C.4 Practice Test Answer to Chap. 4 Problem

Practice Test Answer: Chapter 5 (see. Fig. C.5)

Chapter 5 Practice Test answer

NUMBER OF DAYS SPENT IN THE NEST

NORTH REGION	SOUTH REGION					
95	90	Group		n	Mean	STDEV
98	92	1 North		15	104.67	6.53
99	94	2 South		13	97.69	4.53
112	93					
103	95					
105	98	Null hypothesis:	μ_1		=	μ_2
106	100					
109	102	Research hypothesis:	μ_1		$\neq$	μ_2
110	103					
111	104					
112	98	1/n1 + 1/n2				0.14
114	102					
95	99					
98		(n1 - 1) x S1 squared				597.33
103						
		(n2 - 1) x S2 squared				246.77
		n1 + n2 - 2				26
		s.e.				2.16
		critical t				2.056
		t-test				3.23

Result: Since the absolute value of 3.23 is greater than the critical t of 2.056, we reject the null hypothesis and accept the research hypothesis.

Conclusion: Hedgehogs hibernated for significantly more days in the North Region than in the South Region (105 days vs. 98 days).

Fig. C.5 Practice Test Answer to Chap. 5 Problem

Practice Test Answer: Chapter 6 (see. Fig. C.6)

Research question:	"Does the distance from the nearest recycling center affect the frequency of visiting that recycling center?

DISTANCE (km)	FREQUENCY (days/year)
1	55
2	25
2.5	35
4	30
5	20
6	25
4	28
7	15
7.5	12
5.5	17
4.5	26
1.5	40
2.5	30
3.5	35

SUMMARY OUTPUT

Regression Statistics	
Multiple R	0.85
R Square	0.7272
Adjusted R Square	0.7044
Standard Error	6.0521
Observations	14

ANOVA

	df	SS	MS	F	Significance F
Regression	1	1171.3918	1171.3918	31.9807	0.0001
Residual	12	439.5367	36.6281		
Total	13	1610.9286			

	Coefficients	Standard Error	t Stat	P-value	Lower 95%	Upper 95%	Lower 95.0%	Upper 95.0%
Intercept	47.09	3.7322	12.6179	0.0000	38.9607	55.2242	38.9607	55.2242
X Variable 1	-4.71	0.8334	-5.6551	0.0001	-6.5291	-2.8973	-6.5291	-2.8973

Fig. C.6 Practice Test Answer to Chap. 6 Problem

Practice Test Answer: Chapter 6: (continued)

(d) a = y-intercept = 47.09

  b = slope = −4.71 (note the negative sign!)

(e) Y = a + b X

  Y = 47.09 − 4.71 X

(f) r = correlation = −.85 (note the negative sign!)

(g) Y = 47.09 − 4.71 (4)

  Y = 47.09 − 18.84

  Y = 28.25 days/year

(h) About 19–21 days/year

Practice Test Answer: Chapter 7 (see. Fig. C.7)

TOTAL GALLONS USED TO DRIVE FROM ST. LOUIS TO INDIANAPOLIS

FOUR-DOOR SEDANS

TOTAL GALLONS USED	WEIGHT (1000 lbs)	HORSEPOWER
6.1	3.8	130
6.3	3.7	150
4.8	4.0	140
4.2	2.4	125
3.8	2.9	98
4.7	3.0	115
3.5	2.1	121
5.5	2.9	123
5.9	3.1	110
3.4	2.1	96

SUMMARY OUTPUT

Regression Statistics	
Multiple R	0.77
R Square	0.593
Adjusted R Square	0.477
Standard Error	0.787
Observations	10

ANOVA

	df	SS	MS	F	Significance F
Regression	2	6.320	3.160	5.102	0.043
Residual	7	4.336	0.619		
Total	9	10.656			

	Coefficients	Standard Error	t Stat	P-value	Lower 95%
Intercept	0.29	1.877	0.154	0.882	-4.150
WEIGHT (1000 lbs)	1.01	0.509	1.984	0.088	-0.194
HORSEPOWER	0.01	0.020	0.614	0.559	-0.035

	TOTAL GALLONS USED	WEIGHT (1000 lbs)	HORSEPOWER
TOTAL GALLONS USED	1		
WEIGHT (1000 lbs)	0.76	1	
HORSEPOWER	0.60	0.65	1

Fig. C.7   Practice Test Answer to Chap. 7 Problem

Practice Test Answer: Chapter 7 (continued)

1. $R_{xy} = .77$
2. $a = $ y-intercept $= 0.29$
3. $b_1 = 1.01$
4. $b_2 = 0.01$
5. $Y = a + b_1 X_1 + b_2 X_2$
   $Y = 0.29 + 1.01 X_1 + 0.01 X_2$
6. $Y = 0.29 + 1.01 (3.8) + 0.01 (126)$
   $Y = 0.29 + 3.84 + 1.26$
   $Y = 5.39$ gallons
7. $+.76$
8. $+.60$
9. $+.65$
10. The better predictor of TOTAL GALLONS USED was WEIGHT with a correlation of $+.76$.
11. The two predictors combined predict TOTAL GALLONS USED only slightly better ($R_{xy} = .77$) than the better single predictor by itself.

Practice Test Answer: Chapter 8 (see. Fig. C.8)

NO₂ CONCENTRATION (ppb) IN THE EXHAUST FUMES OF VEHICLES

CAR	MINIBUS	BUS
63	46	38
65	48	40
71	47	41
72	49	43
69	50	44
63	52	46
65	53	47
66	56	42
67	58	48
71	54	37
70	57	36
68		38
		39
		40

Anova: Single Factor

SUMMARY

Groups	Count	Sum	Average	Variance
CAR	12	810	67.50	9.91
MINIBUS	11	570	51.82	17.16
BUS	14	579	41.36	14.40

ANOVA

Source of Variation	SS	df	MS	F	P-value	F crit
Between Groups	4436.04	2	2218.02	161.19	4.49E-18	3.28
Within Groups	467.85	34	13.76			
Total	4903.89	36				

CAR vs. BUS

1/n CAR + 1/n BUS	0.15
s.e. CAR vs. BUS	1.46
ANOVA t-test	17.91

Fig. C.8 Practice Test Answer to Chap. 8 Problem

(b) H_0: $\mu_1 = \mu_2 = \mu_3$

   H_1: $\mu_1 \neq \mu_2 \neq \mu_3$

(f) $MS_b = 2218.02$ and $MS_w = 13.76$

(g) $F = 161.19$

(h) Mean of CARS $= 67.50$ and Mean of BUSES $= 41.36$

(j) critical $F = 3.28$

(k) Result: Since 161.19 is greater than 3.28, we reject the null hypothesis and accept the research hypothesis.

(l) Conclusion: There was a significant difference in NO_2 concentration between the three types of vehicles.

(m) H_0: $\mu_1 = \mu_3$

   H_1: $\mu_1 \neq \mu_3$

(n) df $= n_{TOTAL} - k = 37 - 3 = 34$

(o) $1/12 + 1/14 = 0.08 + 0.07 = 0.15$

   s.e $=$ SQRT $(13.76 * 0.154)$

   s.e. $=$ SQRT (2.12)

   s.e. $= 1.46$

(p) ANOVA t $= (67.50 - 41.36)/1.46 = 17.90$

(q) critical t $= 2.032$

(r) Result: Since the absolute value of 17.90 is greater than the critical t of 2.032, we reject the null hypothesis and accept the research hypothesis.

(s) Conclusion: CARS had a significantly higher level of NO_2 than BUSES (67.50 ppb vs. 41.36 ppb).

Appendix D: Statistical Formulas

Mean
$$\overline{X} = \frac{\Sigma X}{n}$$

Standard Deviation
$$\text{STDEV} = S = \sqrt{\frac{\Sigma(X-\overline{X})^2}{n-1}}$$

Standard error of the mean
$$\text{s.e.} = S_{\overline{X}} = \frac{S}{\sqrt{n}}$$

Confidence interval about the mean
$$\overline{X} \pm t\, S_{\overline{X}},$$

$$\text{where } S_{\overline{X}} = \frac{S}{\sqrt{n}}$$

One-group t-test
$$t = \frac{\overline{X}-\mu}{S_{\overline{X}}}$$

$$\text{where } S_{\overline{X}} = \frac{S}{\sqrt{n}}$$

Two-group t-test

(a) when both groups have a sample size greater than 30

$$t = \frac{\overline{X}_1 - \overline{X}_2}{S_{\overline{X}_1 - \overline{X}_2}}$$

$$\text{where } S_{\overline{X}_1 - \overline{X}_2} = \sqrt{\frac{S_1^2}{n_1} + \frac{S_2^2}{n_2}}$$

$$\text{and where } df = n_1 + n_2 - 2$$

(b) when one or both groups have a sample size less than 30

$$t = \frac{\overline{X}_1 - \overline{X}_2}{S_{\overline{X}_1 - \overline{X}_2}}$$

$$\text{where } S_{\overline{X}_1 - \overline{X}_2} = \sqrt{\frac{(n_1-1)S_1^2 + (n_2-1)S_2^2}{n_1+n_2-2}\left(\frac{1}{n_1} + \frac{1}{n_2}\right)}$$

$$\text{and where } df = n_1 + n_2 - 2$$

Correlation
$$r = \frac{\frac{1}{n-1}\Sigma(X-\overline{X})(Y-\overline{Y})}{S_x\, S_y}$$

where S_x = standard deviation of X
and S_y = standard deviation of Y

Simple linear regression $Y = a + b\,X$

where a = y-intercept and b = slope of the line

Multiple regression equation $Y = a + b_1\,X_1 + b_2\,X_2 + b_3\,X_3 +$ etc.

where a = y-intercept

One-way ANOVA F-test $F = MS_b/MS_w$

ANOVA t-test $ANOVA\ t = \frac{\overline{X}_1 - \overline{X}_2}{s.e._{ANOVA}}$

where $s.e._{ANOVA} = \sqrt{MS_w\left(\frac{1}{n_1} + \frac{1}{n_2}\right)}$

and $df = n_{TOTAL} - k$

where $n_{TOTAL} = n_1 + n_2 + n_3 +$ etc.

and where k = the number of groups

Appendix E: t-Table

Critical t-values needed for rejection of the null hypothesis (see Fig. E.1)

Fig. E.1 Critical t-values
Needed for Rejection of the
Null Hypothesis

sample size n	degrees of freedom df	critical t
10	9	2.262
11	10	2.228
12	11	2.201
13	12	2.179
14	13	2.160
15	14	2.145
16	15	2.131
17	16	2.120
18	17	2.110
19	18	2.101
20	19	2.093
21	20	2.086
22	21	2.080
23	22	2.074
24	23	2.069
25	24	2.064
26	25	2.060
27	26	2.056
28	27	2.052
29	28	2.048
30	29	2.045
31	30	2.042
32	31	2.040
33	32	2.037
34	33	2.035
35	34	2.032
36	35	2.030
37	36	2.028
38	37	2.026
39	38	2.024
40	39	2.023
infinity	infinity	1.960

Index

© Springer Nature Switzerland AG 2021
T. J. Quirk et al., *Excel 2019 for Environmental Sciences Statistics*,
Excel for Statistics, https://doi.org/10.1007/978-3-030-66277-6

Printed in the United States
by Baker & Taylor Publisher Services